Theodore L. Mead

The Life and Times of
THEODORE MEAD

Orchids
and
Butterflies

PAUL BUTLER

ORCHIDS AND BUTTERFLIES: THE LIFE AND TIMES OF THEODORE MEAD

Copyright © 2016 by Paul Butler

Published by Little Red Hen Press.

All rights reserved. This book may not be reproduced in whole or in part, in any form (beyond copying permitted by Sections 107 and 108 of the United States Copyright Law, and except limited excerpts by reviewer for the public press), without written permission from Paul Butler. For permission, or to contact Paul Butler, please email casajard@gmail.com.

For a complete list of photo credits, please see *Picture Credits* at the end of the book.

Author services by Pedernales Publishing, LLC.
www.pedernalespublishing.com

Cover design: Paul Butler and Jose Ramirez

Library of Congress Control Number: 2016913615

ISBN 978-0-9979666-6-4 Hardcover Edition
ISBN 978-0-9979666-8-8 Paperback Edition
ISBN 978-0-9979666-7-1 Digital Edition

Printed in the United States of America

Dedication

To the memory of Kenneth F. Murrah, of Winter Park, whose portrayal of Theodore Mead in historical reenactments became the inspiration for this book.

Epigraph

"The greatest service which can be rendered any country is to add a useful plant to its culture."

                         Thomas Jefferson,
                         Memorandum of Services to My Country,
                         September 2, 1800.

Contents

Preface	*xi*

Part One – The Early Years

1 When Samuel Met Mary, 1843–1845	2
2 Hudson Valley Upbringing, 1845–1861	8
3 Schooling in Germany & America, 1861–1867	14
4 The Grand European Tour, 1867	21

Part Two – Butterflies Flutter By

5 First Visit to Florida, 1869	28
6 With W. H. Edwards in West Virginia, 1869–1870	36
7 Chasing Butterflies in Colorado, 1871	45
8 New Species or Darwinian Variant? 1872–1874	54

Part Three – University Days at Cornell

9 Secret Societies & the Delta Upsilon Fraternity, 1874–1875	62
10 Fraternity Conflicts, 1876–1877	70
11 Escaping Botheration out West, 1878	76
12 Alpha Delta Phi Brotherhood, 1878–1881	82

Part Four – A New Life in Florida

13 Florida Land Purchases, 1881	90
14 Marriage to Edith Edwards, 1882	98
15 Honeymoon in England, 1882	105
16 The Eustis Years, 1883–1886	109
17 Dr. Henry Foster, 1885–1886	118

Part Five – The Move to Lake Charm

    18 New Beginnings & Birth of a Daughter, 1886–1887   *128*

    19 Improving Lake Charm & Bringing up Dorothy, 1888   *137*

    20 Bumper Citrus Harvests, 1889–1891   *144*

    21 Scarlet Fever Strikes, 1892   *152*

    22 Irrigation & The Great Freeze, 1893–1895   *160*

Part Six – Everyday Life in Central Florida

    23 Humanitarian Efforts in the Community, 1895–1897   *172*

    24 Financial Problems, 1897–1901   *179*

    25 Acceptance of Faith, 1903   *187*

    26 Medical Issues & Loss of Family Members, 1902–1916   *194*

Part Seven – The Master of Horticulture

    27 Orchid Breeding, 1891–1925   *206*

    28 Citrus Freeze Protection, 1899–1905   *214*

    29 Caladium Growing with Henry Nehrling, 1885–1920   *221*

    30 Amaryllis, 1889–1930   *229*

    31 Bromeliads, Crinum & Other Flowering Bulbs, 1885–1936   *237*

Part Eight – The Final Years

    32 The Oviedo Boys & Scoutmaster Mead, 1920–1930   *246*

    33 Modern Conveniences & Medfly Crisis, 1925–1930   *257*

    34 Awards, Recognition & Culmination, 1927–1936   *266*

    35 Aftermath & Legacy, 1936–1940   *284*

Acknowledgments   *291*

Notes   *294*

Picture Credits   *352*

Preface

It was on a warm, dry day in late November 2009 that I first set foot in Mead Botanical Garden, entering via the small and almost hidden pedestrian entrance on South Pennsylvania Avenue in Winter Park, Florida. Recently arrived from England as an immigrant and with forty years of horticultural experience, I was keen to discover how time had treated the Garden that when it opened in 1940 was proclaimed as "Florida's Finest Garden Spot." I was in for a considerable shock.

The passage of almost seventy years had not been kind to the Garden. I searched in vain for any signs of orchids, bromeliads or caladiums, plants Mead had been famous for, or any unusual flowering plants or shrubs of botanical interest. Eventually at the west end of the Garden, close to Denning Drive, I came across a large collection of camellias, but that appeared to be it as far as any botanical component was concerned. Close by, a sad, dilapidated and empty greenhouse, open to the elements through its torn and tattered roof, greeted visitors. A sense of neglect hung over the Garden like an unwashed shroud.

Several weeks later, I discovered a small and dedicated band of community-spirited people under the banner of the Friends of Mead Garden, trying to slow the deterioration and turn the Garden around. It was at one of their small and intimate soirées, designed to publicize the Garden and engage the community in its restoration, that I first encountered Mr. Theodore Mead, in the form of a well-known Winter Park resident, Kenneth Murrah, performing a short but captivating historical reenactment of Mead as part of the evening's entertainment. I left with the impression of just having met a fascinating person from the past who had lived in interesting times. Just what was his story, I wondered, and how did this legacy Garden come about?

A minor piece of Internet research revealed that the archives of Rollins College contained thousands of Mead's letters, cataloged in over thirty boxes. The second discovery was of more than a hundred of his glass-plate negatives in the Winter Park Library archives. With these two primary sources but a stone's throw from where I lived, it was as if his story was begging to be told.

Although Central Florida documented most of Mead's life, trips to Brooklyn and Ithaca in New York, to the Florissant Fossil Beds Monument of Colorado, and to Coalburg in the Kanawha Valley of West Virginia were necessary to complete the picture. In particular, the ancestral home at Coalburg of his wife, Edith Edwards Mead, was a treasure trove of photographs and manuscripts. The information in the letters there allowed the last twenty years of his life to be accurately documented, and the collection contained a wonderful selection of photographs, many of them identified and dated.

For Mead's story, a "life and times" narrative seemed particularly appropriate. His lifetime achievements in entomology and horticulture were substantial and significant and are celebrated in this volume. As are his generosity of spirit and concern for others less fortunate than himself that endeared him to the communities in which he lived. The times in which he lived—the environmental, technological and cultural changes that took place during his eighty-year life—provide a fascinating backdrop to the narrative. He experienced the birth of photography and became an enthusiastic early adopter; he witnessed the harrowing death of his only child from scarlet fever, the scourge of juvenile mortality with at that time no known cure; and he became embroiled in the heated religious arguments that followed the publication of Darwin's *On the Origin of Species*. In this debate, from his own studies of environmentally-induced changes in butterfly variants, he took the side of Darwin and that brought him head-to-head in conflict with his evangelical mother.

An early issue of writing his biography was what to call him? In letters, his mother always addressed him as "Theodore," his father as "Ted," his brother as "Theo," his wife as "Teddy," and his family and young friends as "Uncle Teddy." Plain "Mead" seemed too formal, so "Theo" ended up being the compromise.

This then is an account of the life and times of Theodore Luqueer Mead—who he was, how he lived and what made him tick.

Paul Butler
Winter Park
February 2016

Part One

The Early Years

CHAPTER 1

When Samuel Met Mary, 1843–1845

At twenty-five minutes to six on the morning of Monday, February 23, 1852, Mary Cornelia Mead gave birth to her second son, Theodore Luqueer Mead, in the small rural hamlet of Fishkill, sixty miles up the Hudson River from New York. Her first son, Samuel Holmes Mead Jr. (Sammy), had been born there in 1848; now he had a younger brother to play with. Theodore was born before the doctor and nurse arrived, and delivery was easy as Mary described in her diary of that day:

> The Lord was with me, and strengthened me, and dealt very mercifully with me. Had comparatively but little pain and it was of short duration. The Lord heard my prayer. He safely delivered me of a perfect, a proper, a living male child. O that my Theodore may indeed be filled with the Holy Ghost even from his infancy!

Theo's father, Samuel, had met Mary when they were both twenty, and she became the girl next door. Samuel was working in his father's wholesale grocery business at Coenties Slip in the Lower East Side and living with his two unmarried sisters on Second Avenue, at No. 110. Sometime in 1843, the Luqueer family had purchased the row house next door, No. 112, and the two had met formally on New Year's Day, 1844. Mary recalled the occasion in a letter she sent to Samuel in December 1849, "This will probably reach you New Year's Day. This is a memorable day for me. Well do I remember this day six years ago, when a certain gentleman, I will not mention whom, called at Pa's in the evening and inquired for Miss Mary."

1.1: *The Meads and the Luqueers were wealthy New York families, owning successful businesses and extensive properties there. Theo's paternal grandfather, Ralph Mead, founded one of the city's oldest wholesale grocery stores, while his maternal grandfather, Francis Luqueer, established a flourishing saddlery and harness business.*

Mary was the eldest daughter of Francis T. Luqueer, proprietor of a successful harness and saddlery business on Murray Street in New York. She attended Miss McClenachan's Boarding & Day School for Young Ladies at 13 Carroll Place, off Bleecker Street, a step away from her home at the time on Thompson Street. At school, she proved to be an intelligent and exemplary student, collecting numerous certificates of achievement, and was particularly keen on languages, learning

French and German to a standard where she could compose two or three-page letters. She was also a passionate member of the Dutch Reformed Church and a child of the Second Great Awakening, being converted at an early age to a life of evangelical submission to God. As a result, she saw her role on Earth as helping to restore righteousness to a world drenched in wickedness, and this would include the conversion and religious upbringing of her children.

By contrast, there is some evidence that Samuel Mead was a bit of a rebel as a youth and an independent thinker. His mother, Sarah Mead, was a devout Methodist and the writer of her memorial in 1873 reveals the struggle she had with one of her children's religious orientation. This wayward son could only be Samuel, and the parental solution was to send him to Wesleyan University, then a strictly Methodist institution. Much like his ancestors before him, Samuel had problems receiving education in such an evangelical atmosphere, with the result that he dropped out after his junior term.

1.2: Samuel's mother, Sarah Holmes Mead, was a devout and pious Methodist.

The match between Mary and Samuel was therefore potentially incompatible. Mary had given herself to God, and this lifelong commitment was at the center of her life and dictated all that she did. There would be no compromise in following the commands of the Lord. For Samuel, the religious position of living a life of complete holiness, strictly following the Gospel in everything he thought and did, was not necessary to be a good Christian.

In a letter from rural Greenwich in 1844, and as their relationship deepened, Samuel began to share his dreams of escaping into the countryside and living in their "cottage by the woods away from that disgusting city." He concluded, "We shall exchange the din of carts and cabs and stages for the gentle and harmonious sounds of country life; I at least shall exchange the suffocating stench of the docks for the breath of flowers and the perfume of hay."

While Samuel was dreaming about the scent of hay, Mary was wrestling with her religious conscience over her future marriage commitment to him. It was "agitating her soul"— was there room in her life for both the Lord and him? If only Samuel were a man of the cloth, things would be different. Over the summer of 1844, she wrote to Samuel, choosing this form of communication where she could calmly compose her thoughts.

In the letter, she started by expressing surprise that he "never proposed to labor in the vineyard of the Lord." After a few more paragraphs, she delivered the key question on her mind, "Will you not then be a co-worker with God in winning souls to Christ? This is the wish of my heart. When thinking of you I cannot express to you my longings of desire that you should become a minister of the Gospel."

Samuel's reply was firm and carefully crafted and was a rerun of the arguments he had had with his mother. He pointed out to Mary that it was wrong to think "God is to be served only in the pulpit, that there is no station of usefulness except that of the preacher of the Gospel." He went on to state his own personal view that religion was most effective with acts and not words, and he believed that "the unobtrusive labors of a pious mother with her children are as pleasing to God as the solemn worship of the sanctuary." Shifting gears somewhat, he protested that his oratory skills were feeble and his theological knowledge sketchy at best, so he would be ill-adapted anyway to the job of a preacher. "Why spoil a good merchant to make a bad preacher?" he wanted to know.

His concluding comments must have melted away any further resistance:

> I cannot but thank you for your touching and eloquent letter. Some parts of it have affected me to tears my dear, dear girl, I felt myself better and happier for loving you. You seem to me like my guardian angel. I shall pray more in future because your praying will go up with mine to the throne of our Father. I shall care more for the things of religion for the thought of even being separated from you is dreadful.

1.3: Marriage portraits of Samuel Holmes Mead and Mary Cornelia Luqueer, dated 1845 when they were both twenty-two years old.

Mary's love for Samuel overcame her desire for him to be as committed to Christ as she was. "I thank my Heavenly Father for the love of one who cares for my soul," she wrote to Samuel, and they were married on July 2, 1845, in the North Dutch Reformed Church at the corner of Fulton and William Streets in New

York, in a ceremony officiated by Pastor Reverend John Knox. As was the tradition in this age before photography, and for society weddings of this kind, the commemoration of the appearance of bride and groom on this special day was frequently immortalized in oil paintings, done as a pair with the groom on the left and the bride on the right.

After their marriage, Samuel got his dream of living in the country. A house and considerable sized garden, orchard, farmland and outbuildings in Fishkill came as part of the settlement when Samuel's father, Ralph Mead, married his second wife, Anne van Wyck of Fishkill, in 1845. This allowed them to move out of their respective Second Avenue New York homes and replace the bustle of city living with a life of pastoral tranquility.

CHAPTER 2

Hudson Valley Upbringing, 1845–1861

Supported by a considerable inheritance from both sides of the family, Samuel became a hobby farmer, keeping chickens, growing vegetables such as peas and potatoes for the table, and hiring part-time labor at hay time and when harvesting grain crops such as oats. He carefully documented in notebooks his farm expenses and credits, and this discipline of record keeping he passed onto his children, which later would stand Theo in good stead for documenting future scientific and horticultural activities. Samuel appeared to have inherited the "back to the land" instincts of his ancestors who, except for his grandfather, were all successful Connecticut farmers.

At an early age and encouraged by his father, Theo began to take an interest in the garden and the growing of plants, and it is likely that his lifelong interest in horticulture started there. From the age of four, he was often with his father tending the garden. By six, he had his own area, which he later recalled with great fondness, "My own flower garden was my chief delight, mostly annuals but with the crowning glory of one big amaryllis—the fore-runner of the tens of thousands raised in later life."

At the age of five, his love of the natural world was confirmed. The seed of a twenty-year-long passion for collecting butterflies and moths was triggered by a garden find which had him running into the house to show somebody his discovery. He recalled his early admiration for a lovely caterpillar "as big as anybody's thumb and glorious with ruby and turquoise decorations—afterwards known as the Cecropia—and my keen regret as the unsympathetic nursemaid tossed it into the fire as a noxious worm."

2.1: *Theo's initial interest in collecting butterflies and moths started with the chance discovery of a caterpillar of North America's largest native moth, the Cecropia Moth* (Hyalophora cecropia).

An early decision was to tutor the two boys at home where Mary could apply the necessary religious inputs. Lessons on weekdays involved spelling and arithmetic in the mornings, and Bible readings and Heidelberg Catechism lessons in question and answer format in the afternoons. Sunday was a day devoted to the Lord—church attendance was three times during the day and any other activity hinting at enjoyment or pleasure forbidden. His mother frowned on calling on friends and family but permitted reading the Bible or any other educational texts.

Mary found it hard to instruct two exuberant and intelligent boys, and her diary reflects the internal battle going on between her love for them and her insistence that they obey without question her religious teaching and beliefs. Having lost the fight for her husband's soul, Mary was determined not to repeat the defeat with her children. Her frustration manifested itself in the application of regular corporal punishment to the boys, which she often afterward regretted and prayed for help to find a better solution. For his part, Samuel always reassured his wife that he loved and valued her, and in keeping with the romanticism of the Victorian age, wrote many verses of poetry to her, filled with clichés and easy rhyme.

Homeschooling seemed to be working better for Theo than for Sammy. A spell at the local school for Sammy was not a success either, so something had to be done. In 1857, at age nine, his parents sent him away to boarding school in Morristown, New Jersey. With Mary also away attending her sick father, a more relaxed atmosphere for learning existed at home for Theo, and Samuel wrote to Mary that "Theodore is as good as he need be, gives no trouble and seems to enjoy himself. He has got through five times in his multiplication table and learns his verse regularly." While she was away, Mary's response focused on what was important to her, "Does dear Theodore report his Bible lesson and Catechism thoroughly every day? Does my dear husband enjoy constant communion with his Saviour?"

Theo's father was very indulgent and let him do almost anything he wanted so long as it was of an improving and educational nature. As a result Theo read widely and voraciously, choosing books of an adult factual nature rather than children's books, as he later recalled, "I remember at the age of six the fascination of a little comprehended manual of chemistry—I would take it to my favorite perch near the top of a cherry tree and devour its fairy tales of changing combinations which seemed as good as anything by Hans Andersen."

The downside to homeschooling for Theo, with his elder brother away at boarding school for much of the time, was the lack of social contact with playmates of his age. His principal companions on the farm were the family cats and pet rabbits, and his father noted in letters to Mary how attached Old Diddy the adult cat was to Theo, following him around so much that "he complains that he cannot dig in his garden, on account of her getting in the way of his digger." After the death of Old Diddy, a neighbor gave Theo a little black kitten, very plump and playful, which his father again noted, "The little thing follows him over the farm and everywhere and actually climbed an apple tree to get near him."

During winter, Theo skated on the farm pond when the ice was smooth and thick, but other athletic activities were unknown to him. As he stated in his autobiographical article in the American Amaryllis Society Yearbook, "I never had a baseball in my hand until over 22 years of age." It is likely that these unfulfilled social childhood aspects shaped his character and were an important factor in channeling his interests throughout adult life towards boy-centric friendships and activities.

The problems with Sammy's academic performance continued and there was a move to a new school in Poughkeepsie in 1860. However, his report cards there were no better, and his parents sent him a letter dated April 2, with their individual thoughts on his sub-par performance. Mary put it this way:

> My dear Son, I feel grieved by the bad reports that have been received from you. If you consult your own case and gratify your selfishness in its varied forms, you will sink lower and lower in the scale of being, and become an outcast to society, and even worse lose your soul. How sad to contemplate the present results of all our love, painstaking, and self-denial for you! What are you doing to honor, to show your love for us?

His father's response to the problem, written on the other side of the same letter page, carried the message a little differently:

> Dear Sammy, Mamma was obliged to leave for New York before finishing her letter. What she says is very forcible. You should remember that you are now training for your future welfare and if you pass the seed-time you cannot expect any harvest. Your last report places you at zero which is not what might be expected from a lad of your intelligence and capacity.

2.2: Photographs of Samuel and Mary, taken sometime in the 1850s when they were living in Fishkill. Both parents were dedicated to the welfare and happiness of their children, although their approach to their upbringing was very different.

Sammy, like his younger brother, was intelligent but just too independently minded to conform behaviorally in a structured classroom of mixed ability children. A third school switch, to Elie Charlier's French Institute at 48 East Twenty-fourth Street in New York, swiftly followed in late 1860, with Sammy staying with his maternal grandmother on Second Avenue. Unfortunately, the feedback from the principal, Elie Charlier, was familiarly negative:

> I am not satisfied with Samuel, he needs to be repeated the same things too often, he forgets himself very easily. His French teacher, this afternoon, complained of his laughing at his classmates when making mistakes in their reading lesson. … It is the same thing with many other secondary trifles, but which, being constantly repeated, acquire some importance, and besides, the ocean is formed of drops. We must train the best colt when young.

Enough was enough for his parents whose one thought and object of their lives was the welfare and happiness of their children. They decided on a complete change in educational environment and looked to Europe to provide it, specifically to Germany, generally regarded in the United States at that time as having the best and most disciplined educational system. Mary's inheritance settlement following the death of her father was available to fund an extended study period there, and over the summer of 1861 they hatched a scheme to go to Europe as a family for several years. The outbreak of the American Civil War in April, and the thought that in time their eldest son, Sammy, might be dragged into the conflict, almost certainly contributed to this decision.

The plan was for the family to sell their Fishkill farm and move into a rented property at 34 West Twenty-eighth Street in New York. Sammy would then leave the Institute and go with his father to Europe, where he would undergo private tutoring across a broad range of subjects, including French and German. Meanwhile, Theo would complete his schooling at the Institute, taking Sammy's place until May 1862, and then sail to Europe with his mother to join the rest of the family. At this point, they would identify a suitable international school in Germany and enroll both boys.

The location in Germany almost chose itself—the university town of Heidelberg, on the east bank of the river Rhine, a renowned center of learning and a city rich in schools. It housed the oldest university in Germany, founded in 1386, with a motto (in English) that would have appealed to both parents, "The book of learning is always open." There were other advantages there too. Samuel was a strong believer in the health benefits of hydrotherapy, and the area provided many nearby spa towns where he could take the waters. For Mary, Heidelberg ticked all the boxes; it was a sacred place at the heart of her religious beliefs, being the birthplace of her catechism.

CHAPTER 3

Schooling in Germany & America, 1861–1867

On November 16, 1861, Sammy left the French Institute for the last time. By December 6, he and his father had landed at Liverpool in England from where they made their way to Paris, staying initially at the Hotel du Louvre. Despite the adverse behavioral reports, Sammy had acquired a respectable level of French from his time at the Charlier Institute, which impressed and delighted his father. He wrote to Mary that Sammy "is quite a young man in look and manner and acquiring the requisite 'savoir faire' with great readiness. He went into a milliner's shop with a gentleman who could not speak French and translated things well." To support this view he sent her Sammy's latest carte de visite, taken in Paris just before Christmas.

Meanwhile back in New York, in contrast to his brother, Theo brought a far more studious application to his lessons at the Charlier Institute and received good report cards and a number of certificates of merit. However, his overenthusiastic use of the gymnasium in March caused him to break his arm, which lost him a month's worth of study. On May 17 he had completed his term of study and left

3.1: Theo's brother Sammy aged 13 in Paris.

school, aged ten, en route with his mother to Europe, where the family was reunited. They moved into a lodging house in Heidelberg in the Grand Duchy of Baden, under the care of Thomas Koster. Samuel and Mary arranged private tutoring for the boys in the classics, natural history, and the sciences, and language classes for all of them in French and German.

With lessons over for the summer, the family traveled first to Baden-Baden to take the waters and then on to Homburg, which they thought a most attractive location with an excellent and popular spa. As one reference guide noted, "The pleasantest arrangement is when the water is served out fresh, as it issues from the open wells at Homburg, by active, good-natured girls."

At Homburg, English and American visitors, believing in the healing properties of the waters, almost outnumbered the local German population. The Meads stayed there for five weeks, and Mary recorded their experiences in her diary of June 1863:

> We left Heidelberg for Homburg. Daily lessons in German were given us by Mr. Becker. Wanted to experience German family life, Mr. Becker introduced us to the church councilor of Gonzenheim, a small village about one mile from Homburg. Samuel was ordered by Dr. Friedlieb whom we consulted to take baths and to drink the Elizabethan waters every morning taking twenty minutes exercise, which has been attended to daily with but few omissions. He appears to be deriving great benefit from the cure. We

have walked frequently in the grounds and gardens of the Kursaal and have often listened to the very fine German music which is played three times a day, out-of-doors in pleasant weather and in the salon when it is wet.

They made a number of American friends, and when the subject of schooling came up in conversations, a small international school in Frankfurt run by Mr. Kirchhoffer was recommended. After seeing it and meeting the teachers, the parents enrolled Sammy and Theo there for a year starting in the late summer of 1863.

The Frankfurt school, consisting mainly of German boys with a sprinkling of American, English, French, Dutch, Spanish and Italian students, provided a classical education with strong natural history and science components. Hours were 7 a.m. to 5 p.m. in summer and 8 a.m. to 6 p.m. in winter, with two hours intermission at dinner time and fifteen minutes for play between each hour of recitation or study, so neither of the boys believed they were particularly overworked. "The fare was abundant but of the most unearthly and intensely German character, and we had five meals per day," reported Theo in a letter to a friend. School finished at three o'clock on Wednesday and Saturday afternoons and in the summertime the students would often take long walks into the country accompanied by teachers, "often stopping at a country beer garden and regaling ourselves from the least even to the greatest on beer, black bread and cheese."

The school provided regular lessons in Theo's favorite topic, Natural History, which confirmed his love of the subject, especially entomology, a favorite collecting hobby of his school friends as he later recalled, "practically all the boys were ardent collectors—some only of stamps but most of butterflies and beetles as well."

Theo was growing up fast, as his father wrote in 1863 to a family friend, "Theodore is quite self-reliant and manly so that although he is but little more than eleven years, he passes for much more than that. He is quite confident that he could go to New York from here alone if it was necessary, and I have no doubt he could."

With the boys in the boarding school in Frankfurt, Samuel and Mary were so fond of Homburg that they decided to stay there and take lodgings in the small village of Gonzenheim just outside Homburg. There was an express train service between Frankfurt and Homburg, a distance of only eleven miles, and the boys traveled almost every weekend to be with their parents.

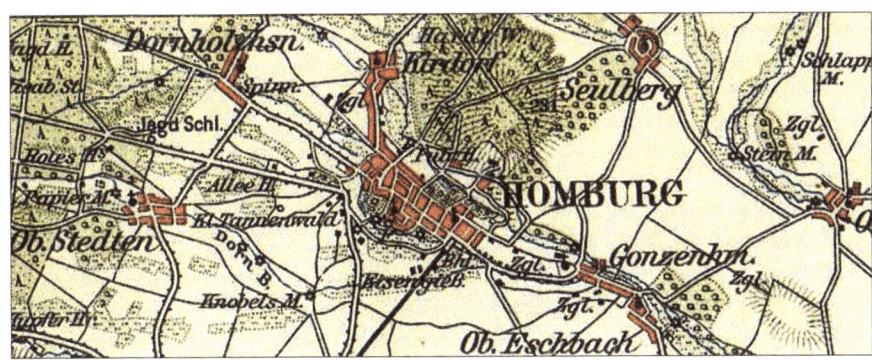

3.2: When the boys started school in Frankfurt, their parents moved to the village of Gonzenheim, at that time just outside the town of Homburg.

During the week, in her letters to the boys, Mary kept up the religious sermonizing, "Prove that you love your parents by trying to please us always. If you have the fear of God ever before your eyes, then you will not fail to give us joy," went one; in another, "I fear that you are sadly neglecting your duties of prayer and reading your Bible! Come anew to the dear Saviour. He will forgive and bless you." She also chastised her sons for not going to church with the other boys at the school, to which Theo replied, tongue-in-cheek, "Respecting going to church, if I must go with the other boys I should also like to go and drink 'bier' with them."

Theo's enthusiasm for his lessons in chemistry and natural history became all-consuming, and he quickly became bored with some of the other subjects taught, such as piano, singing, fencing and physical exercise. He expressed his displeasure to his mother:

> Today is exactly like all the others, a monotonous succeeding of lessons and pauses. This evening some of the boys are playing chess, some are playing checkers with a paper marked with a lead pencil and with beans for men. … Excuse the shortness of the letter and the horribleness of the writing. For the first I offer an excuse that in this most monotonous of all

German schools I have not anything to say. For the second I trust to your great good nature hoping that I will see you in health.

3.3: Theo's parents in Germany, circa 1860s.

The school year finished in the summer of 1864, and they made plans to return to America from the German port of Bremerhaven on the *SS New York* of the

North German Lloyd line. Traveling in upper saloon class, the family of S. H. Mead (40), Mary Mead (40), Samuel (16) and Theodore (12) arrived in New York on August 29, 1864.

By September, the family was living with Theo's grandfather, Ralph Mead, at 233 West Thirty-fourth Street in New York. Both boys attended public schools in the city with Theo preparing for entrance to the College of the City of New York. After the high standards of German tuition and language skills they had acquired, neither was happy with the quality of teaching nor the lack of any general intellectual challenge. They found the education consisted of mostly parroting recitation by rote, and the teachers themselves, mostly Tammany political appointees, unable to pronounce many unusual words. For Theo, as far as subjects were concerned, "work in chemistry or any branch of Natural History was just play."

At that time there were no organized athletics among the public schoolboys—not even baseball teams—and corporal punishment was still the primary motivator, dealt out for the slightest misdemeanor by a severe caning with a rattan in the hands of the principal. Theo hated seeing this happen to boys less fortunate than him, as he later recalled:

> They never offered to give me a licking; I think they were rather respecters of persons and my parents would have removed me from the school rather than permit any assault and battery on me. But some of them got thrashed and to see it done hurt almost as badly as if I had been beaten myself, although none of my particular friends were involved.

On the morning of July 23, 1866, Theo's seventy-seven-year-old grandfather, Ralph Mead, left their house on his way to visit Central Park. On crossing Sixth Avenue, he was struck and run over by a carriage and died later of his injuries. He was buried in the Green-Wood Cemetery in Brooklyn, where he was one of the original plot owners.

In early 1867, Theo entered the Introductory or Sub-Freshman class of the College of the City of New York. He later remembered his class positions, "My rating in Science was about No. 35 and in Latin, etc., about No. 60 in a class of 500 students." But he was soon bored, and dissatisfaction with his locality and schooling brought on a strong wanderlust. The family was wealthy, and he was independent and an experienced traveler, so he saw no obstacles in another trip to Europe, where he could immerse himself in different cultural experiences and use his newly acquired French and German language skills.

His target was the French International Exposition to be held that year in Paris. The rationale became the significant educational exposure provided by forty-two nations exhibiting their most notable works of art, industrial inventions, and developments. Theo approached his father first who agreed if he could find another dependable boy to go with. Theo found a couple of good friends who were interested, but their parents backed out, so his mother offered herself as chaperon "so I should not be disappointed."

CHAPTER 4

The Grand European Tour, 1867

Theo and his mother started from New York in mid-March 1867, and after a fifteen-day trip across the Atlantic on the US Mail steamer *SS Fulton* they arrived at Le Havre, France. Fog held them up leaving the New York Bay area, and the resultant crossing was rough with Theo suffering from seasickness for several days. From Le Havre, an express train took them to Paris.

They spent about a week in Paris, mainly at the French International Exposition, which formally opened on April 1 and closed on October 31, 1867. With over 50,000 exhibitors and almost 10 million visitors, it was the greatest of all international expositions up to that time. The fifteen-year-old Theo admitted later to having been overwhelmed and a little confused by the sheer number of exhibits, but was drawn and kept coming back to the mechanical ones, particularly the display of Jacquard looms weaving damask and beautifully figured silks. After a side trip to Versailles and Sevres, where they bought some medallions of Napoleon III, they journeyed to Naples and Vesuvius, then Rome.

Theo was lonely for male company of his age, but at the Hotel Minerva he met a young Dutch boy and they explored Rome and the surrounding area together. After a week, his friend's vacation time ended, leaving Theo feeling downcast. He wrote to his brother, "Yesterday the only young friend I had in Rome quitted this city for Florence, leaving me alone in my glory, as far as young people are concerned. It is a position I don't fancy much. He was a Dutchman and told me that if I should ever be up his way (Rotterdam) I must be sure and knock at his door."

Theo and his mother took an apartment in Rome and stayed there until the end of May. It was Holy Week in Rome when they were there, which suited Mary but did not impress Theo, describing it as a "grand humbug." They took regular Italian lessons, with mixed results as Theo described in a letter to Sammy, "You would laugh tremendously if you heard Mamma trying to speak Italian. She makes in a sentence of sixteen words, about three English, five French, seven German and one Italian. As to my own efforts I shall maintain a profound silence!"

Their travels saw them in Florence on May 26, Turin on June 1, Como and the Italian Lakes on June 9, and Venice on June 13. In letters written to his father and brother back home, Theo provided plenty of vivid detail of their time there:

> We went home and took the 12:40 train for Venice. We reached this City of the Lion of St. Marks, this city of crabs and shellfish, this groggy, wet and nasty marinely-smelling city of magnificent palaces and former seat of tremendous power (comparatively), about 9 o'clock in the evening and paid a man a whole franc for taking our trunk upstairs! The gondolas are very cheap—1 franc (18¢) for the first hour. We are in the Hotel Barbesi and are quite comfortably situated, 18 francs per day for both 'tutto compreso' as the Italians say. This morning we took a gondola and visited two private palaces and then the Palace of the Doges. In front of it stands the famous Lion of St. Marks. He seems to have white china eyes and has a tail that looks as if a poker had been straightened and attached to the statue; only that the handle-end had been dipped in dross so as to leave an irregular bunch of ragged metal at the end.

At the end of June, they traveled to Austria then east to Adelsberg in Slovenia, site of the famous cave containing eyeless fishes and described later by one travel writer as being like "Milan cathedral turned upside down." From there they stopped at the mercury mines of Idrija, at Cracow to visit the thousand-year-old salt mines, and then on in mid-July to Warsaw, Moscow, and St. Petersburg.

4.1: In Dresden, Theo persuaded his mother to buy him a butterfly collection from the shop of Otto Staudinger, a leading dealer in Lepidoptera.

Theo called St Petersburg "without exception the most magnificent city I have ever seen, at least the buildings are." In the palm greenhouse of the Botanical Garden, the oldest one in Russia founded in 1714 by order of Peter the Great, he was amazed at the sight of one of the tallest coconut palms pushing its fronds through the glass dome, even though they had already dug a hole forty feet deep for its root system. From St Petersburg, their route took them by sea to Helsingfors (now Helsinki) and on to Stockholm, then to Christians (now Oslo) and Copenhagen. They arrived in Germany at Hamburg on August 18, then on to Hannover, Göttingen and Berlin, seeing various German friends they had met a few years earlier in the Heidelberg and Frankfurt area.

By then his mother was exhausted and ready to go home. In Dresden, Theo came across a sale catalog from a shop owned by Otto Staudinger, a leading authority and specialist in butterflies and their collection. Enthralled by what he saw, Theo begged his mother to buy him a large and comprehensive collection of butterflies priced at $50. His homesick mother agreed on the condition that they omitted Holland and Belgium and went straight to Paris for a few days and then home. But Theo wasn't finished, and would only concede if his mother would make it $50 in gold—gold being at a premium at this time. His mother was happy to agree to the deal; she was going home and here, at last, was a specific interest that her son could channel his energy and interests into when they returned home.

After a week in Paris in mid-September, they crossed the channel to England. Theo was amused by the customs inspection on entry, writing, "The baggage was 'examined' at Dover, the examination consisting of asking us if we had any cologne which being answered in the negative we were passed." Theo had written home from Paris regarding their transatlantic passage booked on the Cunarder *Tripoli* leaving Liverpool on September 28, and their probable arrival date home.

The trip had taken almost seven months and embarking on this European grand tour of religious and natural history attractions was breathtakingly ambitious, even by today's standards, amounting to more than 5,000 miles, excluding the transatlantic crossing. Back in New York in mid-October, there was time for Theo to reflect on the trip. Since their time at the French Exposition, his mother had heavily biased the itinerary towards the sightseeing of buildings and objects of a religious nature, with a few isolated exceptions involving Natural History. Theo had enjoyed the exceptions but wrote about his overall experiences:

My mother was the most conscientious person in the world and the improvement of my mind lay heavily on that "New England conscience"— all too heavily I thought, as I plodded my weary way through not less than a hundred miles of mostly mediaeval paintings and sculptures all the way from south Italy to Central Russia and Scandinavia and pretty much everything in between. The botanical gardens and museums were the bright spots though I was rather intrigued by Galileo's dried forefinger in Florence and also by the stuffed skin of a saint.

Part Two

Butterflies Flutter By

CHAPTER 5

First Visit to Florida, 1869

Back in New York from Europe, Theo devoured all the reading matter he could get his hands on relating to butterflies. Butterfly collecting now became his passion and the search for new and unusual specimens became a central part of his life.

After a course of tutoring, he followed his elder brother by entering the School of Mines at Columbia College, but once again found school dull and boring after the excitement of the Grand European Tour. In December 1868, aged sixteen, he was elected the youngest member of the American Entomological Society of Philadelphia and punctually attended their monthly meetings, traveling from his home in New York. In that year, the Society published the first volume of William Henry Edwards' monumental work *The Butterflies of North America*, and Theo naturally acquired a copy and quickly became familiar with the detailed contents.

While a student at Columbia University, his mother and Sammy made a trip with friends to Florida over the winter of 1868. Their letters and descriptions were so irresistible that Theo nagged his father until he agreed to let him join them.

Theo left New York in late February 1869 and sailed down the east coast of America in the side-wheel steamer *Champion* of the New York & Charleston Steamship Line to Charleston. There he switched to the steamer *Dictator* that operated a semi-weekly service to Palatka, Florida, with intermediate stops at Savannah, Fernandina, and Jacksonville, where he was reunited with his mother and brother. After a week there, they departed on the morning of Saturday, March 13 on board the *Darlington*, bound for Enterprise on the north shore of Lake Monroe. The 205-mile trip took 36 hours and the fare was $9.

5.1: The earliest known photograph (tintype) of Theo, taken on February 14, 1868, a few days before his 16th birthday.

5.2: Whitney's Florida Pathfinder colored map of the upper St. Johns River from Palatka to Enterprise, dated 1876. Lake Harney and Salt Lake, where the Mead brothers had fun shooting alligators, are also shown.

Once past Palatka and Lake George, the upper stretches of the river meandered and narrowed considerably. With sharp bends, the boat had to wind its way beneath overhanging branches of shoreline trees, requiring skilled navigation. Crew members had to be on the lookout for snakes that sometimes slithered down onto the boat deck from the Spanish moss draping the overhanging trees, and for alligators that might get tangled up in the paddlewheels. Shooting from the upper decks of the boat was actively encouraged to prevent this, provide sport for male passengers, and help stave off boredom on the long journey. A deckhand helped by pointing out the alligators hiding in the shoreline vegetation, and frequently men lined the deck and shot at anything that moved. They gunned down herons, egrets, ospreys, buzzards and other birds noted for their beauty, as often as the hated alligators and snakes. Sometimes if a large alligator was killed it would be roped and hauled aboard to be gaped at, then butchered for the tail meat and its head cut off so that the teeth could be removed. These were much prized by the tourists, who considered them to be attractive ornaments when carved and polished.

The result was indiscriminate carnage on a scale that would horrify us today, although this was acceptable behavior then at a time of almost complete disregard for the environment. There were few dissenting voices; one of the most notable being Harriet Beecher Stowe, author of *Uncle Tom's Cabin* (1852), who wrote in 1873 of her disgust at the practice in Florida:

> Alas, there is only one sorrow, one blot on the pleasure (of boating). The slaughter of the innocents! On the deck of the boat are men who see no beauty in nature, who have no sympathy with the wild, free, lovely life of the forest, and whose only aim is to leave a bullet in every palpitating living creature they pass. … What makes the thing more inexcusable is that there is no pretense of game. Nothing shot is taken, or pretended to be taken, and half of the men are bad marksmen that can only mutilate.

5.3: To pass the time, male passengers frequently shot at alligators, birds and wild game from the top deck of St. Johns River steamboats.

Sammy and Theo blasted away with the rest of them from a prime position on top of the pilothouse of the boat, hitting five alligators and killing two, as recorded in Sammy's diary entry for March 15.

Arriving at Enterprise, they stayed at the Brock House. While their mother relaxed and read, the boys went off exploring. Camping out in the wilderness never seemed to bother Theo or his brother—it was all part of the excitement. Reassurance against wild animal attack came in the form of the firearms that the two of them carried. But whereas Theo had trouble hitting a barn door at twenty paces, Sammy was an expert marksman.

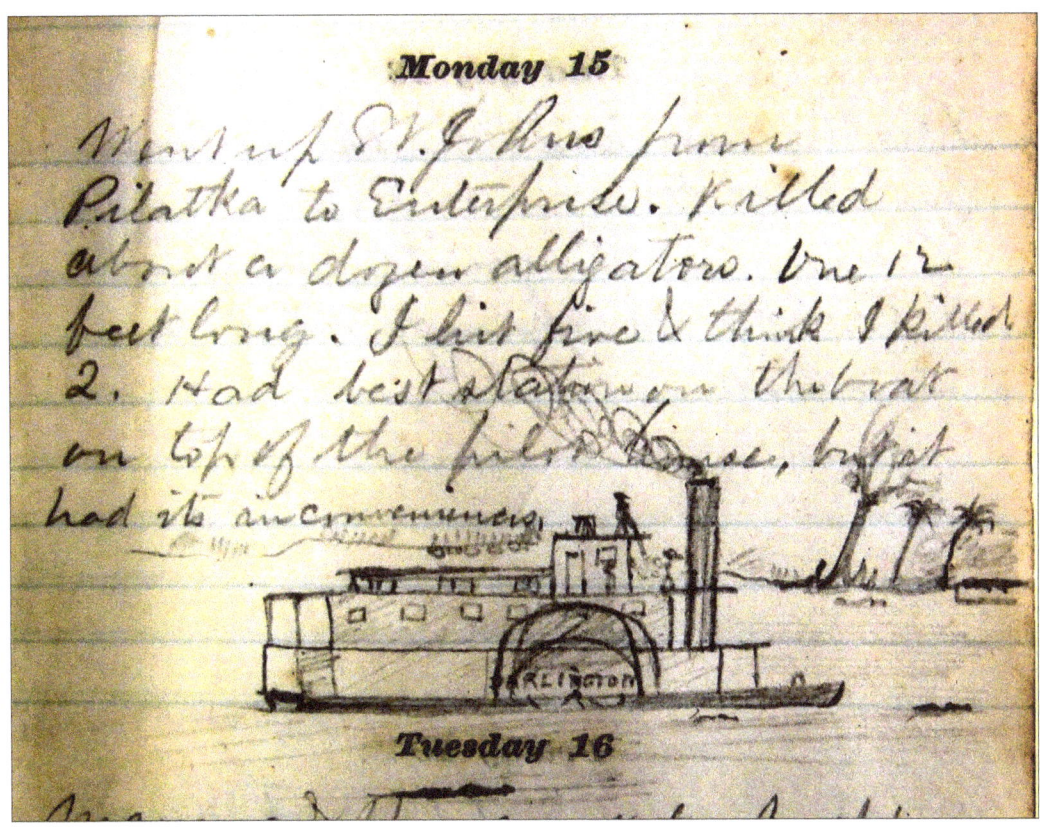

5.4: Sammy's diary entry for Monday, March 15, 1869, recording the shooting of alligators.

One day Sammy shot a fish hawk (osprey) on the wing with his gun and was astonished to measure the bird's wingspan at 5 feet 4 inches. Although his rifle was very accurate, it caused terror in bystanders when discharged as it scattered powder grains in all directions.

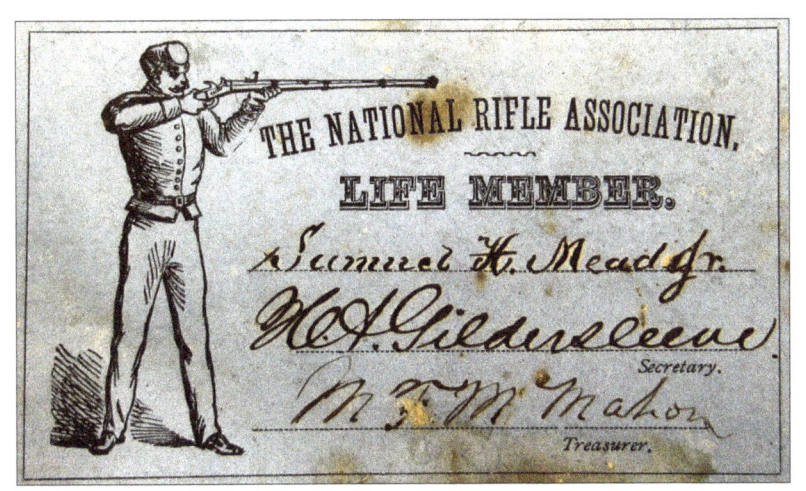

5.5: Sammy was a crack shot with a rifle and a member of the National Rifle Association.

First Visit to Florida, 1869

One day they all took the steamer *Hattie* up to Lake Harney and found alligators in abundance, reporting seeing fifty-four going up with Sammy hitting nine and killing three, the largest about twelve feet long. They wrote home that "The river water was bad, in fact, alligator soup," and referred to the gators as their "friends," whose "features are not handsome but have an openness about them when they smile." A week later the two boys rented a rowboat and repeated the trip to Lake Harney. They intended to explore further upstream, as Sammy informed a friend, "Tomorrow Theo and I put our boat and a week's rations on the *Hattie* for Lake Harney forty miles up. Thence we shall sail and row towards Salt Lake, camp out, slay certain alligators and come home again in our boat."

Up to now, Theo had built up his butterfly collection by exchanging and purchasing species with other collectors. The real thrill for him was to be out in some wilderness area and come across and capture a rare and beautiful new species. This visit to the Enterprise area gave him such a thrill when in April he managed to net a female of *Papilio calverleyi*, a spectacularly rare specimen. A single male, captured in August 1863 on Long Island, New York, was the only other example of that type ever found before.

5.6: Papilio calverleyi *as illustrated in W. H. Edwards' seminal reference work,* The Butterflies of North America, *Volume 2.*

Theo had his butterfly and Sammy had shot just about everything that moved, so all things considered, they had had a very successful time and resolved to return to Florida someday. It was time to go home, but not before the two of them had experienced New Orleans, which they reached by way of Jacksonville and Mobile. After several days in the city, they boarded the riverboat *Mary Houston* bound for Cincinnati via the Mississippi and Ohio rivers, thence by rail back to New York via Washington.

CHAPTER 6

With W.H. Edwards in West Virginia, 1869–1870

After the triumph of his Enterprise catch, Theo had no difficulty in plucking up courage to write to the author of the book he most coveted, and the foremost authority on butterflies in North America, W. H. Edwards of Coalburg, West Virginia. He wrote and asked him to recommend a suitable place to collect within a radius of 500 miles of New York. Mr. Edwards replied that where he lived in West Virginia along the banks of the Kanawha River would be a good spot, and he could come down and spend the summer with him if he liked. Theo seized the chance to learn from the great man, obtained boarding rooms locally, and was with him all that summer.

In 1869, Coalburg was a remote place, inaccessible from the east other than by a long and arduous stage journey over the Virginias. The conventional route and the one Theo took was by rail to Baltimore, then via Harper's Ferry to Parkersburg on the Ohio River, and thence by steamboat to Gallipolis, near the mouth of the Kanawha, where a second boat conveyed passengers up the Kanawha River to Charleston and on to Coalburg.

Because of the isolation, the Edwards saw few people from the North and visitors to the house were restricted to a few ornithologists and entomologists wishing to consult and study with him.

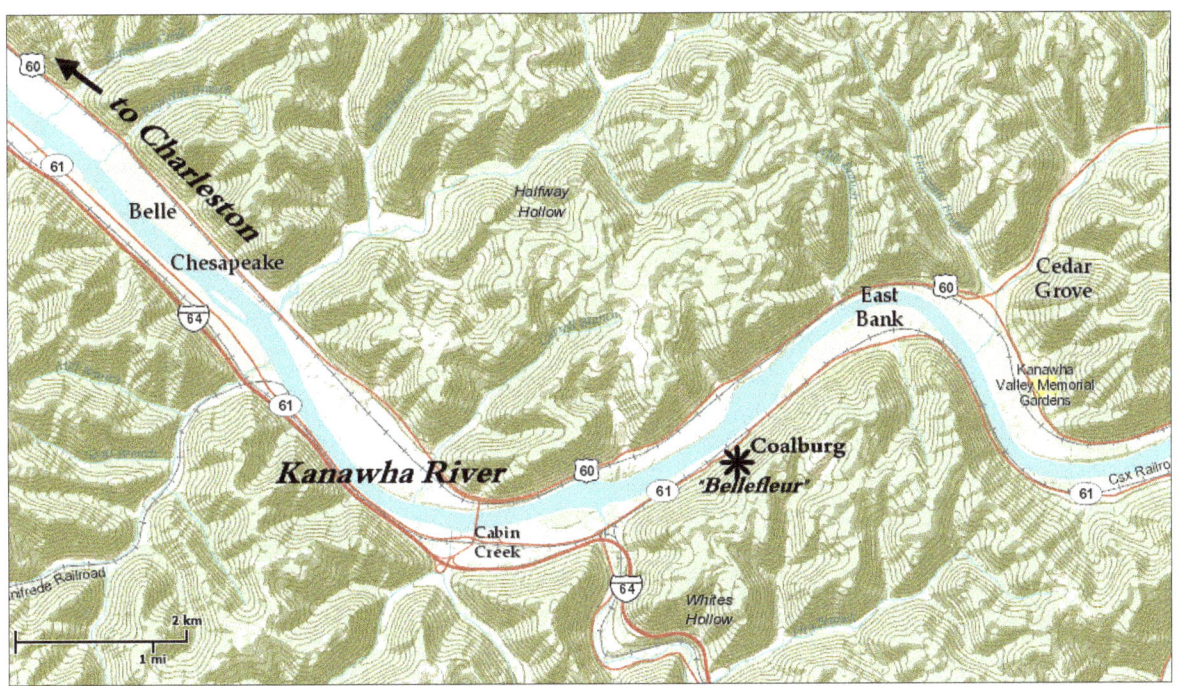

6.1: W. H. Edwards lived at Coalburg, east of Charleston, West Virginia, on the south bank of the Kanawha River. The historic family home Bellefleur *still exists today, marked on this modern map.*

Born in the Catskill Mountains at Hunter, New York, in 1822, W. H. Edwards became interested in natural history as a boy. As a young man, he devoted his attention to ornithology, assembling an impressive collection of local stuffed birds from the Hunter and Catskill areas. In 1847, he joined a party to explore the Amazonian area of Brazil, where he collected many birds for his collection and subsequently wrote an account of the trip in the book *Voyage up the River Amazon*, which was widely read at the time.

6.2: Theo at his desk in New York in June 1869, studying a butterfly before his first visit to Mr. W. H. Edwards in West Virginia.

Eventually, ornithology gave way to entomology and then specifically to lepidopterology. Around 1856 at New Hamburg, New York, he found many sorts of caterpillars and Mrs. Edwards made colored portraits of them. In 1859, the family moved to Newburgh, on the Hudson, where they lived for the next ten years. There his collection increased considerably and via the acquaintance of other lepidopterists, he received many species and developed an extensive knowledge of their localities and habits.

On the death of his younger brother in 1845, Edwards had inherited thirty thousand acres of land in West Virginia, bought cheaply from a speculator, which became the basis of his future business in coal and oil exploration. He opened the first coal mines on Paint Creek in 1852 and erected the first cannel coal oil works in 1856. In August 1864, shortly before the end of the American Civil War, he founded the settlement of Coalburg, on the south bank of the Kanawha River, where he operated extensive coal mines for the Kanawha and Ohio Coal Company of which he was president. He built a house there in 1869 and the rest of the family joined him.

6.3: *William Henry Edwards, around 1880, the foremost authority on North American butterflies at the time.*

The Coalburg area proved to be a goldmine for new butterfly species, particularly the region at the mouth of Paint Creek, five miles up-river. There he observed a male *Speyeria diana* (Diana Fritillary), at that time one of the rarest American species in collections. A few days later he took a female, the first time one had been seen and captured. But the workload in collecting and documenting the many new species began to eat into his time running his mining business, so Edwards welcomed the help of a hard-working apprentice like Theo.

W. H. Edwards was an agnostic and committed Darwinian, believing that most people made their own way in life and had no need for a sovereign God. That butterfly species evolved, and variations in things like size and color (polymorphism) were the result of environmental factors, seemed obvious to him. He would put this beyond scientific doubt in a series of brilliant experiments in the 1870s and 1880s. Theo greatly respected Edwards as the leading American lepidopterist of the day and shared both his religious and Darwinian views. Still only seventeen and a keen and intelligent learner, Theo quickly became an indispensable apprentice, getting eggs and meticulously recording the life history of the butterflies he collected.

After the summer in West Virginia, Theo returned to the family home on West Thirty-fourth Street, New York, and joined his brother studying at the Columbia School of Mines. Theo and Sammy were close, sharing in each other's interests and scientific and engineering curiosity. By their late teens, they had seen most of Europe and both were fluent in French and German.

While Theo was attracted to the natural world, Sammy became increasingly passionate about astronomy, so much so that his parents bought him a telescope. At the Columbia School of Mines, he developed an interest in the physical properties of the metal and glass of telescope construction and the effect these properties had on the limits of resolution. He wrote several pieces for the *Scientific American*, a New York publication specializing in scientific and engineering advances. In these articles, he discussed telescope resolution, how to control temperature expansion and contraction, and how to make glass lenses of improved optical quality. With his expertise in the development of improved telescopes for astronomical observation, he joined the debate as to whether the United States should fund a "Million Dollar Telescope" to rival the resolving power of the best European ones.

As a keen rifle shooter, Sammy was a member of "The Amateur Rifle Club of New York City," and a life member of the National Rifle Association, taking part in their competitions. He was also an inventor and in 1872 patented a type of explosive bullet for use in hunting, which claimed to stop even the largest grizzly bear in its tracks.

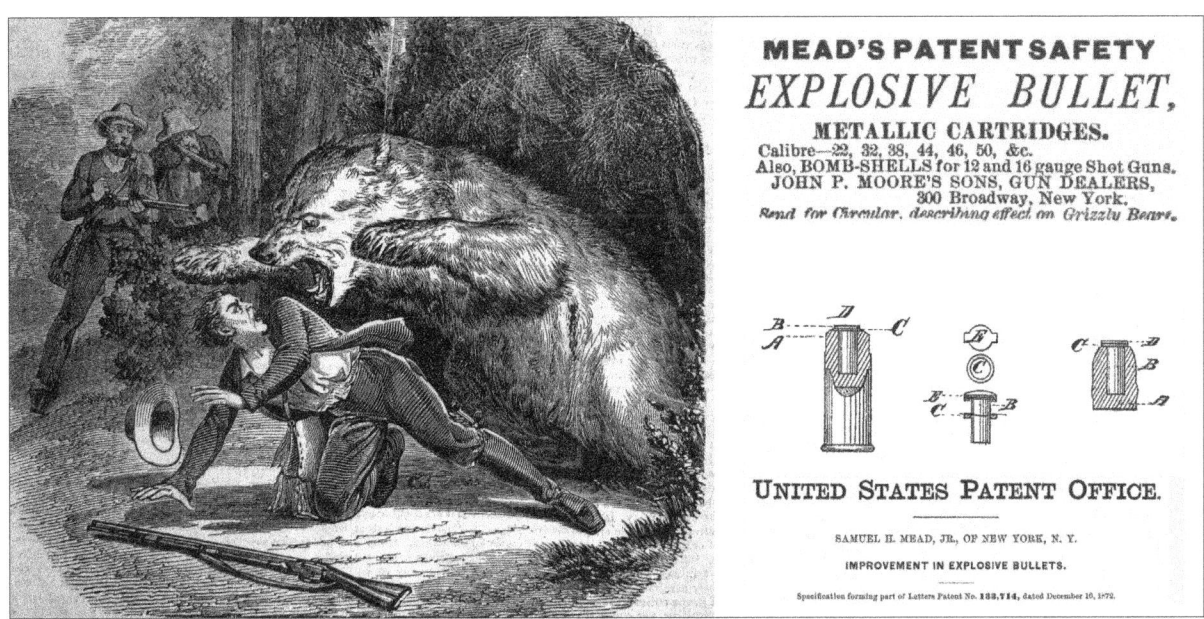

6.4: Theo's brother was an inventor and in 1872 patented a type of explosive bullet for use in big game hunting.

Both brothers had a broad knowledge of chemistry and developed a keen interest in the new technology of photography. They acquired the necessary chemicals to prepare sensitive photographic substrates, and Theo experimented by taking tintype portraits of family and friends.

In early 1870, his parents decided to build a new house on Madison Avenue at the corner of East Sixty-first Street and furnish and decorate it in European style. They wanted to feature the very best in European furniture and luxurious silks, lace, brocades and linen fabrics. With their wealth and fondness for international travel, Paris was the chosen destination and it was agreed that Mary would go there and buy all she needed, shipping the larger items of furniture and bringing the smaller items back with her.

By August, the new house at 596 Madison Avenue was ready and the furniture from France delivered. Living with eighteen-year-old Theo at this address according to the 1870 census records were his father and mother, aged 47, and Sammy, aged 22. In keeping with their well-to-do position, they employed two female Irish domestic servants aged 22 and 24 years old.

6.5: A tintype photograph of his father, taken by Theo on January 21, 1871.

Theo was back in West Virginia for the summer of 1870 and received a warm welcome from the Edwards and their three children, Edith (18), Will (14), and Anne (12). As far as butterflies were concerned, Theo continued to impress his mentor with his growing knowledge and skill in collecting. His ability to observe and record scientifically improved under Mr. Edwards' tutelage. It became a capability that remained with him for the rest of his life and served as an essential element in his future contributions to horticulture. In his entomological reminiscences, Edwards later stated of Theo, "He was in the fields and woods all the day long and never returned without trophies of his net, and without discoveries in the matter of caterpillars or food plants."

The great gap in knowledge for lepidopterists at that time was the visual connection of the life cycle stages of the butterfly, from egg to adult. No one had shown through illustration the continuous metamorphic changes unique to a particular species, so that the identification of an adult butterfly from the caterpillar form alone was filled with uncertainty. Any slight variations and differences in color between two adults caused an argument among collectors, who could not be sure whether the variations were within a single species or represented two different species.

Edwards recognized that a systematic breeding program using eggs from a known species could solve this problem. The breakthrough came when he discovered that he could readily obtain species-specific eggs by confining pregnant females to the food plants preferred by their caterpillar offspring. Edwards' method hinged on identifying the favorite larval food plant for each species and then making sure he had the plant growing locally in his garden or in a purpose-built heated greenhouse. The critical connection for successful egg breeding was thus through horticulture—Theo's area of expertise—and Edwards relied heavily on his knowledge and experience in the discovery and cataloging of the various food plants.

The Chesapeake & Ohio Railway ran past the house between the house and river, and Edwards arranged for the building of a flag stop there, requiring a telegraph request to the local superintendent of the railroad for permission for the trains to stop. This greatly improved travel time for collectors wishing to visit. It was also extensively used both by the family and by Edwards' team of collectors who could send him butterfly material and eggs almost directly to his door. With the food plants identified and growing there, viable eggs could be shipped by rail directly to Coalburg by his network of lepidopterists.

Edwards then placed the eggs on their respective food plants, where in time young caterpillars would emerge and feed on their natural food, protected by fine netting from flies, birds, and other enemies. Following several molting cycles, they entered the chrysalis stage and finally hatched into colorful winged adult insects.

In this way, not only could the entire life cycle of the butterfly be studied, but also perfect undamaged butterflies resulted from the final dramatic metamorphoses. Once food plant identification and culture was in place, Edwards could continue and extend his studies from his home without the time-consuming process and expense of conventional collecting. He wrote, "It was soon found by me that the females of any species of butterfly might be enclosed in a bag over a growing stem of its food plant, or over a plant rooted in a pot, and thenceforth it became easy to get eggs from any part of the United States and from much of Canada, and even France and England."

CHAPTER 7

Chasing Butterflies in Colorado, 1871

In early 1871, Theo received a request from W. H. Edwards to consider joining the Wheeler expedition of that year and butterfly hunt in Colorado. As expected, the prospect of another great adventure for the two brothers to explore uncharted territory "out West" was enthusiastically received. Although it was officially the turn of sections of other Western states, it was the collective view of Edwards and other expert lepidopterists that Colorado would prove to be the most promising for butterflies. They were known to be plentiful there because of the abundance of wildflowers but had not yet been systematically studied, so there was an excellent chance that new species would be discovered. In Edwards' mind, it did not matter when Colorado was explored since the results of the various Wheeler surveys would not be published until the termination of the entire exercise. What mattered most was the caliber of the collector. It was a simple matter for Edwards to nominate Theo, the most competent butterfly collector he knew, to be a private part of the 1871 survey.

The plan was that Theo and his brother would travel independently and not be part of the official Wheeler party. As such, they would be without the protection of the US Cavalry and would have to depend on the sharp shooting skills of Sammy to keep off bears and the occasional unfriendly Indian. Both brothers were mature in attitude and appearance for their ages (19 and 23) and did not mind roughing it when traveling, considering that camping out under the stars was part of the adventure.

Edwards obtained the necessary railroad passes permitting onward travel to Utah and California to confer with lepidopterists there. He also agreed to bear half the expense of the trip to a limit of $500. Edwards was keen to acquire new material and discoveries for his second volume of *The Butterflies of North America*, and the understanding was that all specimens collected would be sent to him in Coalburg for identification and further analysis.

7.1: At the American House hotel in Denver at dinner on June 1, 1871, Theo and his brother dined with Buffalo Bill Cody (right), who wanted to join them on their travels to South Park.

Theo met up with his brother in Chicago and they traveled together arriving in Denver on May 31, where they stayed at the American House hotel. In a letter to his parents, Sammy reported that at dinner the next evening they met "a brass and crockery drummer of good address and appearance," who introduced himself as Buffalo Bill, said he wrote for Bonner's Ledger and wanted to travel to South Park. He was keen to join up with them because he had heard that the Indians there were very troublesome. After listening to his proposal, Sammy concluded, "The offer was pecuniarily a very good one, but his statements of fact being unreliable we declined with thanks."

Theo and his brother explored the area to the southwest of Denver for the next few months, spending time in the mountains north of South Park and along the stage coach route from Denver to Fairplay and Leadville. Their accommodation consisted of a combination of hotels and ranch houses in the settlements and camping out by streams and lakes. Both had their negatives. Colorado was frontier land and when they camped out security against wolf and bear attack came in the form of Sammy's shooting skills. Even so, there were dangers as Theo underplayed in one of his letters to W. H. Edwards, "We start on Monday morning for Fairplay in South Park, by stage, 17 hours. Indians are friendly—they only killed one man last week." At an elevation of close to 10,000 feet above sea level, even in June, temperatures at night dropped to freezing at times. The game was scarce, but the fishing was good. One afternoon Theo caught about a dozen trout weighing from half to one pound but ended up complaining, "We have trout at every meal and I am getting tired of it."

Staying in hotels was warmer, more secure and the table more varied. But many hotel rooms were infested with bedbugs and bathing exposed areas with kerosene before retiring offered only partial protection. At least Theo could appreciate the irony of the situation. As an entomologist, he collected butterflies and other insects during the day, only to be bothered at night by the unwanted attentions of *Cimex lectularis*.

Chasing butterflies at this altitude also had its problems, and frequently during the day, Theo had to rest and catch his breath while the butterfly he was chasing fluttered slowly out of reach. One day they made the ascent of Mt. Lincoln at 14,295 feet, and after a six-hour climb did not reach the summit until seven p.m. Theo wrote to a friend:

The view was beautiful over mountains and snowfields, plains and lakes ... There was the "Mountain of the Holy Cross," one snowfield on whose side ever retains the form of a Latin Cross. Coming down we had a terrible time. My brother became sick, the horses were gone, night came on and a thunderstorm broke over our heads. Between 9 and 10 o'clock we reached the hotel after wading through the Platte River and getting in a number of mining ditches.

7.2: Theo and his brother climbed Mt. Lincoln and admired the view to the west of the Mount of the Holy Cross.

Despite all these setbacks, the butterfly fauna of Colorado did not disappoint. In the summer months of June and July, the abundance of wildflowers that bloomed in the Rocky Mountain countryside ensured it was alive with butterflies, sometimes found in vast numbers, as Theo reported in 1877:

In no place outside of the tropics have I found a better collecting-ground, at least so far as diurnals are concerned, both as to variety of species and number of specimens. This abundance, however, is chiefly noticeable early in the season, as indicated by the number of specimens I was able to secure in the different months—namely, 1,792 in June, 1,483 in July, 607 in August, and only 43 in September. ... Certain diurnals of arctic types positively swarmed on many of the peaks—for example, *Argynnis helena* (Edwards), and lower down several species of *Melitæa, Phyciodes*, and *Argynnis*, were constantly to be found at flowers.

7.3: Theo discovered 28 new butterflies in Colorado; three were named after him as (L to R), Mead's Silverspot, Mead's Sulphur and Mead's Wood Nymph.

On rainy days, he turned his attention to collecting insects other than butterflies, claiming more than 3,800 specimens by the time he finished. He sent regular shipments of papered butterflies back to W. H. Edwards for identification, and twenty-eight species new to science were discovered. Three of these were named in Theo's honor by Edwards—*Argynnis meadii* (Mead's Silverspot, now *Speyeria callippe meadii*), *Colias meadii* (Mead's Sulphur), and *Satyrus meadii* (Mead's Wood-Nymph, now *Cercyonis meadii*).

By September the butterfly season was essentially over, but before he left Colorado, he stopped one evening at the Hartsel Ranch in South Park, where he heard of a strange place just thirty miles away, known to settlers and Indians for its curious rock samples containing delicate insect and leaf fossils and giant petrified tree stumps. Theo's focus was butterflies but with his entomological background, anything with six legs was of interest. On September 13, he wrote to his aunt in New York City telling her of his visit, "So, I hired a horse and went there finding all as represented. I found nearly 20 insects and brought back about 25 lb. of petrified wood. Some of the stumps are 20 feet across. They are in all respects similar to ordinary stumps but converted to stone."

The twenty insect fossils that he found embedded in shale rock fascinated Theo, and as a Darwinian, he realized how important they were in understanding and telling the story of the evolution of insects. He sent all his samples to Edwards, who sent them on to palaeoentomologist Samuel Scudder at Harvard University. Scudder published his findings in 1876, noting Mead as the collector and naming one of the fossilized insects—a type of subterranean termite ant—*Eutermes meadii*. Mead thus became the first person to recognize the importance of the Florissant Fossil Beds and bring it to the attention of the rest of the scientific world.

7.4: *Theodore Mead was the first person to recognize the scientific importance of the Florissant Fossil Beds in Colorado. He discovered the fossilized remains of an extinct termite ant that was named after him,* Eutermes meadii.

The site contains the remains of massive petrified redwood trees that once dominated the Florissant forest, and beds of fossilized shale from an ancient lake deposit. The rock was laid down more than 30 million years ago and contains fossils of trapped insects, plants, fish, microscopic algae, and pollen. Over the latter part of the nineteenth and early part of the twentieth century, numerous expeditions collected tens of thousands of fossils from the site, distributed to museums around the world. The site is protected as the Florissant Fossil Beds National Monument.

From Colorado, Theo and his brother continued westward in the fall, visiting entomologists in California and making a side-trip to Yosemite. All in all, this had been a most successful trip—Theo's skill and determination had produced a vast amount of new butterfly material for W. H. Edwards. Theo had another reason to be satisfied: he had fallen in love with the breathtaking scenery of the American West. In response to a fellow entomologist who asked how this trip compared to previous ones "wandering over Europe," Theo replied, "Of course there are lovely stops and strange scenes (in Europe) but what profiteth a man that he sees 20 miles of pictured saints and 'Holy families' and loseth the sight of the Rocky Mountains and Yosemite?"

He and Sammy returned East by steamer via Panama, arriving back in New York by mid-December, in time for Christmas. They did some collecting in Panama, filled an empty trunk with orchids, and brought back two parrots, an iguana and a horned toad as pets. In keeping with his precise nature, he noted in his diary the following statistics for 1871: horseback travel, 518 miles; railroad travel, 3,795 miles; steamboat, 6,810 miles; stage coach, 952 miles; total travel from place to place during 1871 = 12,075 miles.

Mead's collecting time in Colorado was remarkable for the number and variety of butterflies he discovered, and was important enough to be subject to much recent historical analysis. Using letters in the archives of Rollins College, lepidopterists Grace and F. Martin Brown reconstructed his Colorado itinerary and published it in 1996 as an illustrated booklet. In addition, in 2012, lepidopterist historian John V. Calhoun discovered Mead's 1871 journal, from which his daily activities for that year are now known in considerable detail.

7.5: *Grace and F. Martin Brown reconstructed Theo's time in Colorado from his letters and copybook and published the results as a booklet in 1996.*

8.1: Nineteen-year-old Theo sitting at his butterfly cabinet in New York. His correspondence reveals that this tintype photograph was taken on Feb 27, 1871, with his mother removing the lens cap of the camera for the duration of the several-second exposure.

CHAPTER 8

New Species or Darwinian Variant? 1872–1874

For the next two years Theo was kept hard at it, sorting and collating his finds from Colorado and working with Edwards to produce new material for volume 2 of *The Butterflies of North America*. He was also given the task of writing up the results of the Wheeler expeditions. This turned out to be a mammoth undertaking since he needed to include descriptions, images and notes of all the butterflies he had collected in Colorado and all the other expedition finds made by other biologists for the years 1871–1874. The final publication was a triumph for Theo, consisting of a detailed and important chapter of over fifty pages in one volume of the seven-volume report.

The Edwards family was now living in a new and larger house on the same site at Coalburg, the original dwelling having burned down in February 1871. The house was built at the mouth of a wooded ravine where the land sloped down to the Kanawha River. The garden was an important feature of the new property, being laid out carefully with beds and borders full of old fashioned flowers, roses, and bright annuals to attract the butterflies, with a greenhouse, fruit trees, berry bushes and a vegetable garden. Years later, from the colorful garden flowers came the adopted house name, *Bellefleur*.

The experiments of breeding butterflies from eggs obtained from females in captivity were proving very successful. It was time for Edwards, with Theo's help, to turn his attention to understanding the origin of polymorphism in butterflies, where adults of apparently the same species showed variations in size and color markings, particularly between spring and summer forms.

8.2: On one of Theo's visits in July 1873, he took his camera and photographed the new Edwards house at Coalburg, with Edith sitting on the steps of the veranda.

The publication in 1859 of Charles Darwin's influential book *On the Origin of Species* had sent a shock wave through conventional religious thinking in the Western world. According to Darwin, polymorphism in nature was a result of evolutionary processes that were also responsible for the development of species from other preexisting species over long periods of time. To Calvinists who believed in a world where "all creatures great and small" were fashioned by a supreme being, as biblically described, species were created by God. Any diversity in nature was all God's doing. Dissent was rife on either side of this philosophical rift, and the well-known Harvard naturalist, Louis Agassiz, dismissed Darwin's views, stating, "Evolution was an insult to the wisdom and will of God." To his way of thinking, all butterfly variants should be classified as new species.

Other experts in the butterfly world were similarly polarized over the evolution versus creation question, since the butterfly was perhaps the most potent symbol of God-created beauty. Also, some Christians thought of the metamorphosis from chrysalis to butterfly as an indirect representation of every Christian's hope of resurrection. To be confronted with the possibility that such beauty and symbolism were merely the result of random chance and a series of evolutionary events was too much for many to swallow.

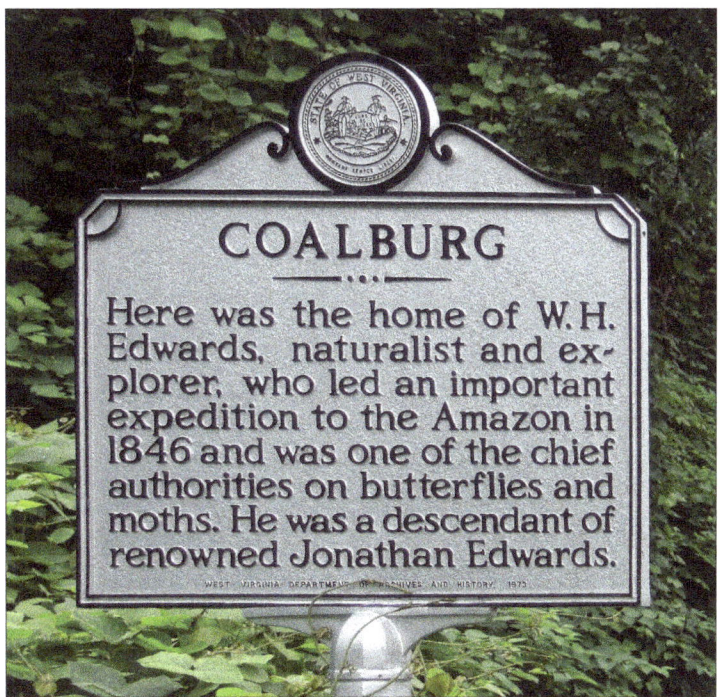

8.3: *The historical marker recognizing W. H. Edwards by the side of County Road 61 on the south bank of the Kanawha River, close to* Bellefleur *in West Virginia.*

With his egg-breeding program, Edwards was able to prove conclusively that from the same batch of eggs, variants could be produced by changing the hatching and growth conditions, season or temperature, showing that the variants were a result of changes in the environment and did not represent distinctly new species.

When Edwards' Volume 2 of *The Butterflies of North America* was completed in 1884, it caused a sensation in the natural history world. In exquisite detail and via superb illustrations, and supported by Theo's seminal work in Colorado, it told for the first time the life-cycle stories of well-known American butterflies. It scientifically explained polymorphism in butterflies and proved that variations within a species originated from environmental factors, directly supporting Darwin's evolutionary thesis. The volume, and subsequently Volume 3, secured Edwards' reputation as the greatest American butterfly expert. It received the accolade as the best book ever written on American butterflies, on the same iconic level as Audubon's bird books.

A feeling of inclusion was essential to Theo, and Coalburg, where he was increasingly spending his summers, became his second home. The natural hospitable and kind nature of the Edwards family ensured that Theo was always warmly welcomed with genuine physical affection from Edwards' son, Will, and the females of the family. He felt comfortable and secure there. The family had fun together, playing parlor games and singing songs around the piano. They were religious but liberally so and had no problem accepting either the agnostic views of the head of their family and his number one apprentice, or the evangelical views of any visitors. On one occasion after a visit, Theo wrote to Sammy, "It is really wonderful what a home atmosphere all the Edwards family carry around with them."

In contrast, life in the Mead household, presided over by their strong-willed and autocratic mother, had a much more formal structure. Religion, knowledge and learning were the honorable past-times. Both Sammy and Theo continued to be at the receiving end of interminable spiritual interrogation from their mother as to their religious activities, either verbally if they were at home or by letter if away somewhere. Occasionally this would drive the boys to distraction, as one letter to Theo from Sammy, still at home in New York revealed, "This month I have commenced to go to concerts and things as it is intolerable to be nagged continuously about piosities and not have society or relaxation." Theo's agnostic leanings and belief in evolution from his work with butterflies must have added additional friction to his relationship with his mother. Like so many devout Christians, she had trouble reconciling evolutionary principles with the belief that the Creator originated all the beauty of the natural world.

Theo spent the summer of 1873 at Coalburg, wrestling with thoughts of his future direction and interspersed with bouts of lazing around and tending to his butterflies. His collection of North American butterflies had grown with his collecting skills and Mr. Edwards' help, until it stood third behind that of Edwards and the British Museum. By this time, it was clear that in terms of butterfly knowledge, Theo had evolved from dedicated apprentice to Edwards' heir apparent, but in his mind he was no means certain he wanted that role. Will Edwards, now aged 18, was set to follow a career in his father's coal and gas mining businesses, and he was not impressed by the monetary rewards of butterfly collecting as an occupation based on his father's experiences. He encouraged Theo to find something more practical to do with his life. Discussions between Theo and Will took place including some light-hearted banter by letter as to what profession he might choose. It was clear that Theo was clueless as to what his future direction might be. In a reply to a letter from Will, he wrote:

> Out of the four professions from which you desire me to make my choice (lawyer, doctor, merchant, thief) the first two and last form a group by themselves being quite similar in nature and in fact the first two are principally modifications of the last being for the most part the latter combined with fibbing and murder respectively. All three involve too much hard work to be attractive. I really haven't the remotest idea of what I ought to be or do.

Will had become his firm friend, and Theo wanted some version of the future that included him. Finally, and somewhat with an air of desperation and not much thought as to the subject on Theo's part, the two of them reached an agreement whereby they would both apply to study at Cornell University in Ithaca, New York, with Will choosing to study science and Theo civil engineering. As Theo later admitted, the choice of his subject was a bit of a mistake, and he regretted not specializing in biology where his real inclination lay. This decision exemplified one of Theo's personal characteristics, that despite being a logical and brilliant scientific thinker, these powers could desert him at critical times and he would end up with a poorly thought-through decision and lingering regrets.

Part Three

University Days at Cornell

CHAPTER 9

Secret Societies & the Delta Upsilon Fraternity, 1874–1875

Theo and Will Edwards arrived at Cornell on October 1, 1873 to enroll Will into the Ithaca Academy, where he was to receive a little more tuition to prepare for the Cornell entrance examination the following year. Theo's general impressions of the steep hills and dramatic gorges of Ithaca were positive, and he wrote to his parents that it was "A very pretty place and I think I will like it." He went back to his parents in New York until January 1874, when he returned and found board and lodgings with Professor and Mrs. W. C. Russel in their house on Seneca Street at $9 a week. Both Theo and Will appeared happy to board there and were not particularly attracted to the Greek fraternities that had arrived at Cornell within a few months after opening in the autumn of 1868.

The establishment of fraternities on campus, with associations of exclusivity, privilege, and secrecy, had initially resulted in the formation of an anti-secret "Independent Organization." This quickly faded from view following intense ridicule directed against its founding members by the established fraternities.

9.1: Theo aged 22, shortly after his arrival at Cornell University in 1874.

In its place, representatives of the Hamilton and Rochester Chapters of the Delta Upsilon fraternity decided to form a Cornell Chapter. This initiative from an anti-secret literary society, founded at Williams College in 1834, released a tirade of abuse and personalized attacks. An 1869 issue of *The Cornell Era*, a university student publication staffed by secret-society members, characterized it as "an association with nothing save its badge to recommend it, a clique utterly anomalous, and without character."

But by the time Theo arrived in 1874, it was the secret Greek fraternities turn to be criticized in a very public row over their initiation practices, triggered the year earlier by probably the first recorded example of death from hazing. Mortimer Leggett, a Kappa Alpha Society pledge at Cornell, was fatally injured when he fell off a cliff on the south side of Six Mile Creek on October 10, 1873, during an initiation ceremony. This event resulted in a public outcry and reaction against the perceived evils of secret societies.

To their supporters, the fraternities built essential social and leadership skills, created networks that lasted a lifetime and provided a close-knit support group within the campus community. The secrets—gestures, handshakes, passwords, mottos, or other knowledge secrets—formed the essential "glue" that bonded the brotherhood members together for life based on loyalty and trust.

Without these bonds, went the argument, the society would be no different from a chess club.

To their detractors, the fraternities were discriminatory, exclusive and self-serving. They created unfair class distinctions where merit was ignored. Fraternity men were generally wealthier and more conservative than many of the other students and their chapter houses were often luxurious, promoting jealousy and hard feelings among the rest of the student body.

9.2: *Theo as a newly elected member of Delta Upsilon, circa 1875.*

Initially, Theo assumed a neutral stance on secrecy in this heated debate, not regarding secrecy *per se* as essential or a particularly desirable element. However, in time he would come to reverse this position, and recognize that the bonds created by secret society membership suited his personality and his need for strong fraternal friendships.

In early March 1874, Theo was invited to join the anti-secret Delta Upsilon Society. This organization, although opposed to secret societies, was never completely open to the public. Attendance at literary exercises was open to all students, but new recruits had to be invited to join and then needed unanimous support to gain membership. Theo attended several meetings and was impressed by the exercises and members. He noted that most of the naturalists belonged, as well as faculty member Professor Russel with whom he boarded. Theo, a natural leader, quickly became the secretary and minute-taker, and started to organize the Society and ensure strict standards in their literary meetings, when each member would take turns in presenting a short oral and verbal dialogue on a subject of their choice.

Theo himself was a prolific presenter on a whole range of subjects, which included essays on *Language as Fossil History, The Use of Anglo-Saxon Words,* and *Method in Daily Life*. He set high standards and could quickly descend into petty preaching, admonishing members for their lateness at literary meetings, or their incorrect use of adjectives rather than adverbs in their written presentations. Some of his essays appeared in the *Cornell Review*. One entitled *Evolution of Sense-Organs,* published in 1875, was a pro-Darwinian treatise that he sent to his mother, producing a predictable response:

> Your piece in the "Cornell Review" (is) sensibly penned as far as it goes, but it is necessary that you should recognize the Divine Artificer in whose hand your breath is and who has made all our numbers fearfully and wonderfully. What inexpressible loss you are suffering my dear child by staying away from Christ! For He hath promised us eternal life if we receive

Him and believe in His Name, make trial of His love and power to save, and you will commence a life of joy and peace, even here, the beginning of everlasting life.

9.3: *Theo at Cornell in late 1875, as a principal fellow of the Delta Upsilon fraternity, wearing a fashionable astrakhan overcoat and matching hat.*

Delta Upsilon was without a Chapter House at this time and had rented rooms at 12 Quarry Street, close to where Theo was now boarding. After a faltering start, fraternal relationships emerged through the natural growth of friendships among the members, and the Society began to prosper. They purchased a piano in 1874 and as well as literary meetings, social events became more frequent. Usually, these consisted of singing college songs, drinking lemonade through straws, telling jokes of a mild and unobtrusive character and generally "having a jolly time." Smoking and beer drinking, even in moderation, were not considered suitable pastimes, and Theo commented in a letter to his brother, "It is a curious fact that of our twenty-eight men none smokes or uses tobacco, a state of affairs one would hardly expect among college students."

At the end of his first year at Cornell, Theo received the 1874–75 President's Prize and $20 for the best lecture on a special subject in physiology, based on his research work with W. H. Edwards on butterflies. On top of that, his Society was flourishing, and all appeared to be going well with the world.

On May 20, 1875, events turned this world upside down. A telegram from his father arrived in the early afternoon at his Ithaca rooms stating, "Brother is ill. Come home. Mailed fifty five (dollars). Telegraph freely." Theo took the first train back to New York to find his mother and father distraught and his brother dead in his bedroom from an accidental gunshot wound to the head.

The situation arose as Sammy was preparing for a trip to Martha's Vineyard with his father, to test out the efficacy of the explosive bullets on the local shark population. He was collecting the last of the equipment and accessories from his bedroom upstairs at 674 Madison Avenue, with his arms full and carrying a hair-trigger rifle in his hand. His father was downstairs when, at five minutes past eleven in the morning, he heard a shot and rushed upstairs to find his son dead. Apparently, the rifle struck a piece of furniture and went off accidentally, the bullet passing through the left side of his temple, killing him instantly. He was twenty-seven.

Mary's diary entry of May 20, 1875, recorded her version of events:

> This memorable day our precious son Samuel Holmes Mead Jr. entered into his glorious rest by a rifle shot passing through his head. The Lord gave and the Lord hath taken away. Blessed be the name of the Lord! The time that God took him was about 11 a.m. We acknowledged God's loving hand in

it. After prayer and some deliberation we sent for his cousin Dr. Frederick E. Hyde, who assisted us to remove him from the closet where he fell, to his bed. There was a coroner's inquest held at about 5 p.m. consisting of Brother Robert, Dr. Hyde, Mr. Bailey, Mr. Van Rodin and his two assistants and they rendered a verdict of accidental shooting.

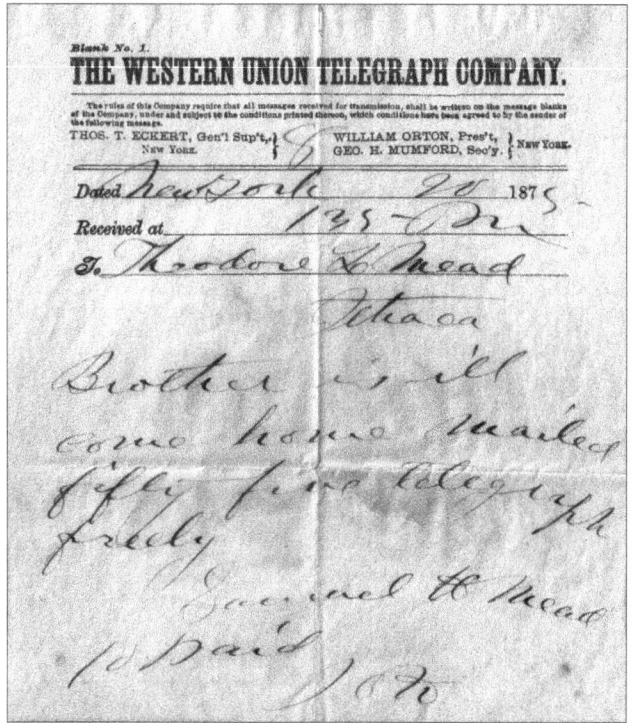

9.4: On May 20, 1875, Theo's father sent a telegram to Theo at Ithaca urging him to come home immediately, where he learned of the death of his elder brother, Sammy.

In June 1875, Theo wrote a long and sad lament for his brother and sent it to the *Scientific American*, a publication that had carried many of Sammy's articles and comments regarding astronomy and the construction of state-of-the-art telescopes. Part of the obituary read, "On the 20th of May while making preparation for a tour of observation and experiment, he died suddenly from the effects of an accidental gunshot wound. Thus, suddenly ended the life of one, whose union of physical vigor and capacity for patient thought and unselfish labor, with rare delicacy of feeling, indicated the strongest type of character."

As the following year progressed, the anniversary of Sammy's tragic death did not go without notice and reflection. Theo's father wrote, "Yesterday I went over to Green-Wood and meditated; it was the anniversary of the darling precious boy's death, May 20th five minutes after 11 a.m.—was there ever such a darling!"

Theo and his brother had been very close, and the loss was devastating. His strong need for close male friendship had suffered a major fracture, and on returning to Cornell he threw himself enthusiastically into society matters and the compensation that brotherly love within the fraternity brought.

CHAPTER 10

Fraternity Conflicts, 1876–1877

From his earliest childhood, Theo had been a cataloger and collector. Images of his fraternity chums at Cornell became one more series to collect, and much like the business cards of today, friends commonly exchanged a carte de visite or cabinet card with their photograph on it. As a result, portrait photographers in a college town like Ithaca did brisk business, and a fashionable gallery was Beardsley & Mackey, known simply as "Beardsley's." Operating as artists rather than photographers, they offered customers the choice between a regular photograph and an idealized portrait, where they retouched the negative to create an image free from blemishes and any shortcomings of the human face. They did business with the advertising testimonial, "You have made me look as I hope to look in Heaven; send two dozen more."

Theo and Will Edwards had their portraits taken at Beardsley's in March 1876, and both were extensively retouched—in Theo's case particularly to strengthen the chin area—and Theo was delighted with the results. He wrote to his parents, "This afternoon I received copies of my pictures from Beardsley—he broke the full face negative so I shall have to sit again for that view. These pictures came out gorgeously and he has made a perfect phoenix of Will. … I enclose a copy of the side face picture from which you can see how much he improves the negatives by retouching."

10.1: Theo was very pleased with this side-face Beardsley portrait of March 1876.

To help get over the loss of his brother, he sought solace in horticulture, and decorated his rooms with cheerful and flowering hanging baskets in every window, filled with Strawberry geranium, *Thunbergia*, Coliseum ivy and *Maurandia erubescens*. At the beginning of his time at Cornell, he had started collecting cacti, writing to his parents, "My cactus plants are all growing nicely and give me great pleasure as well as do other plants," and later, "Some of my cacti are beginning to show flower buds. I have forty-one varieties, including one in the plant case."

His first recorded experiment in plant hybridization dates from around this time, as disclosed to his father:

> I am trying one or two experiments in hybridization now—I have two species of *Abutilon* flowering now; one has variegated leaves and yellow flowers veined with red while the other has pure white flowers. I carefully removed the stamens from a flower of the first before the pollen was ripe and dusted the pistils with pollen of No. 2. The seed capsule is growing nicely and I dare say I may get some quite interesting varieties from the seeds when they ripen.

In May, the society instigated a search for a Chapter house to rent and found a brick twelve-room house in a prominent position for $600 per annum. Ten of the Delta Upsilon Fellows agreed to furnish their rooms themselves and pay $900 towards rent and expenses. Theo was excited about the move and asked his parents for help in providing furniture and carpets. He elected himself in charge of the conservatory to ensure there would be flowers to decorate the parlors at every meeting, and assumed responsibility for keeping the books to give him practice in business affairs.

With this new financial responsibility, Theo suggested that the society rushing planned for the following Fall should focus on "aristocratic fellows" and they should not take in any more "paupers." This resulted in tension within the fraternity membership. One section was keen to broaden the class structure and relax the rigid membership criteria, seeing little wrong with leisure smoking and beer drinking in moderation and the occasional game of billiards. Opposing this position, and in favor of new members being aristocratic in tastes and temperament, were Theo, Will Edwards, and a few others. Since new members needed unanimous approval, quarrels ensued and society life became increasingly unpleasant.

Will Edwards was the most outspoken member of the aristocratic supporters, boycotting any potential candidate whom he thought was of unrefined taste and coarse, whether they were rich or poor. In February 1877, these frustrations reached boiling point and the President of the Delta Upsilon society, William Dudley, called an extraordinary meeting in an effort to expel Edwards. To do this, he required a three-quarters majority vote. One vote short of this target, and determined to remove Edwards, Dudley resorted to chicanery by getting rid of Theo's vote by suspending him for half an hour on some spurious technicality, then expelling Edwards and finally accepting Theo's resignation. Three other Mead/Edwards supporters then also tendered their resignations.

Within a very short time after this acrimonious meeting, all five former members had bids from existing secret societies, who took great pleasure in the results of this dispute and the resultant damage to Delta Upsilon's reputation. By April 1877, Mead and Edwards were members of Alpha Delta Phi, one of the largest and prestigious of the secret Greek fraternities at Cornell.

Academically, Theo had made a creditable record at Cornell despite "much wearisome mathematics," and graduated in the summer of 1877 with a first (Bachelors) degree in Civil Engineering. Not wishing to leave Cornell, and realizing that his true interests lay in natural history rather than civil engineering, and botany in particular, he agreed to start some post-graduate work in the fall of 1877 under Professor Prentiss, Head of Botany.

He spent the long summer break variously with his parents in New York, with W. H. Edwards in the Catskills, and visiting most of the active chapters of the Alpha Delta Phi fraternity in New York and New England, where the brothers received him with affectionate regard. After a spell of butterfly collecting on Nantucket, Theo joined Will Edwards and family at Oak Bluffs, Martha's Vineyard, where sailing, fishing, and bathing were the order of the day.

The essential principle of Alpha Delta Phi membership was "gentlemanly conduct at all times," the perfect social blueprint for Theo. He rapidly found welcoming and sincere friendships and his initiation into Alpha Delta Phi seemed to him the most wonderful and happiest thing that had ever happened. In his autobiographical article in the American Amaryllis Society Yearbook, he would refer to the day of initiation as one of the three wonder days of his life, the others being his wedding day and the day of confirmation in the Episcopal Church.

10.2: The star and crescent fraternity pin of Alpha Delta Phi.

"I now belong to the society which I should have joined when a freshman," he wrote to a friend, and "Now that I have the opportunity of judging between a good secret society and the anti-secret one I can fully appreciate how far superior the former is, not only in things which make life pleasant but also in those things which tend to make the student a man, and worth something in the world."

But by early 1878, despite all the love and support from the fraternity, Theo was still trying to get the "wreck" of Delta Upsilon out of his system and remained in an emotionally charged state, and in letters to his father he described himself as being "intensely unhappy." That his chosen society turned out to be an untrustworthy one in his mind was a great revelation. His father lavished praise on his son's character and stated that the age-old solution to emotional stress of any kind was, in his opinion, a complete change of location, adding, "Let's dodge all botheration and go in for ease and independence. That's the correct card and we can do it as easy as rolling off a log."

He suggested that Theo should take the rest of the year off from Cornell and join the family on a six-month natural history trip to California and the Western States, sweetening the proposed deal by adding that he had heard that Los Angeles was an interesting village and was keen to evaluate property there. As expected Theo jumped at the opportunity and so, on a wintery day in March, the party sailed out of New York harbor, down the east coast of the United States and across the Caribbean, arriving at the Panamanian port of Colon (Aspinwall) on March 28.

CHAPTER 11

Escaping Botheration Out West, 1878

For the Meads, the old Panama route from the East to California had the most potential for exploration. It was longer than the new transcontinental railroad, but there was a greater chance of discovering new butterflies and plants of a semi-tropical nature in areas like Panama and Mexico. All the family was excited at the prospect and Theo ended up describing this trip in detail in his autobiographical article in the American Amaryllis Society Yearbook. At the southern terminus they transferred to the Pacific Mail ship *Colima*, a modern iron-hulled screw steamship of 2,905 tons built for the Pacific Mail Steam Ship Company in 1873. It set sail that evening bound for San Francisco, where the Meads were due to disembark following an extended intermediate stop at Acapulco.

Just before daylight on April 3, they passed a smoking volcano near Libertad in San Salvador, before arriving at Acapulco on April 6, where Captain Coffin anchored in the bay to avoid paying customs duties on the ship's supplies. His father's diary recorded the event, "Next stop Acapulco; we think we can while away two weeks there with pleasure and the next steamer is the *Granada*, a finer vessel than this so we shall not change for the worse. April 6: Arr. at Acapulco about 3 p.m.; we steamed into a beautiful bay wholly land locked and quiet as a pond."

They had a delightful time at Acapulco, making many prospecting trips, but there were no butterflies. They rented a Mexican adobe house and had it and the grounds all to themselves, although the beds were hard and by no means restful. Meals, taken at the residence of the proprietor a short distance away, included mangoes and other fruit from the garden, and for refreshment, they drank the contents of cut green coconuts.

On April 19, they boarded the steamer *Granada* and got underway at one o'clock in the morning, steaming "in full moonlight out of the narrow nick of water to the ocean" on the way to San Francisco. For the last two days at sea, they passed green hillsides with great patches of flowers visible for miles. So much vegetation gave them cause for optimism that this would produce a corresponding abundance of butterflies. They reached San Francisco on April 28 and took accommodation on Market and Montgomery Streets at the Palace Hotel, whose palatial appearance impressed them by being superior to anything they had experienced in New York. The quality of the food served at mealtimes made a much bigger impression. After a month of traveling and more than twenty days eating ship's food, Theo wrote, "Since leaving Acapulco we've had to devour things miscellaneous and various but all odious and evil tasting and as Papa puts it, only swallowed under penalty of death by starvation if it were not done. You haven't any idea how good it is to get something decent to eat again."

After a suitable time to recover and recuperate, they went south on May 7 to butterfly hunt and look at property opportunities in the Los Angeles area. After the previous steamer experiences, they decided to go by train. Theo described Los Angeles as not much more than a village of 16,000 inhabitants, but attractive. Initially, they took a hotel in the middle of town but soon decided to get out into the fields and flowers of the countryside. They headed inland sixteen miles up to the newly opened Sierra Madre Villa, set in the foothills of the San Gabriel Mountains above what is now East Pasadena, where they stayed for a week.

The twenty-room Sierra Madre Villa provided luxury accommodation and spectacular views, positioned in the middle of a beautiful ranch setting and surrounded by vineyards and citrus groves. Guests had running water in each room. The villa had its own bee apiaries and maintained a stable of horses and a small herd of cattle. It grew all its own fruit, vegetables, nuts and grapes which were freshly supplied to the tables three times a day. At that time of year, from the veranda of the property, they could see acres of wildflowers stretching all the way down the valley to Los Angeles.

11.1: The 20-room Sierra Madre Villa was surrounded by orange groves, orchards, and fields of wildflowers.

No doubt seduced by this glorious natural setting, Theo and his father came very close to purchasing a forty-acre plot with an orange grove and house for $6,000. His father said he could have it, but Theo backed out at the last minute. This missed opportunity was a source of much regret to him over the years, and he later stated, "I missed out as Los Angeles now has over a million of inhabitants and petroleum was found all over the place."

They spent days exploring the area on foot and horseback, with both parents helping to net butterflies. Theo wrote about his discoveries to Mr. Edwards and enclosed specimens he could not identify. One day they took a drive to see the ranch of "Lucky" Baldwin, the celebrity businessman, mining investor, racehorse owner and breeder, known not only for his luck but for his fair employment treatment of hired, migrant and immigrant workers. Impressed by the vast size of Baldwin's 63,000-acre ranch, Theo commented, "The Californians seem to have a passion for doing everything on the same scale as the mountains and waterfalls." A week later, it was their turn to stand in awe at these wonders of Nature as they arrived in the Yosemite Valley for a six-week stay, taking them up to the end of June 1878.

11.2: Like the tourists in this period photograph, the Meads explored the trails in Yosemite Valley initially on horseback.

In the valley, part of their stay was at the superbly positioned Snow's Hotel, situated about midway between the Vernal and Nevada waterfalls. From the hotel porches, visitors could experience the falling, roaring water and catch the occasional gusts of mist and water spray if the wind were in the right direction. The excellence of Mrs. Snow's cooking complemented the magnificent location. Her donuts, bread, and elderberry pie drew particular praise, as did her ability to "cook all the popular dishes" including apple pies, hominy cakes, turnovers, and trifles. She also had a dry wit and her standard banter to visitors was to tell them in a serious way that there was always "eleven feet of snow here all summer." When asked how that was possible, she would reply, "My husband is near six feet tall and I'm a little over five. Ain't that eleven?"

The Meads were nothing if not thorough in their exploration. While the vast number of tourists dashed into the Valley and out again, ticking off the major sights, the Meads initially took horses over the many trails but then repeated the trails on foot. Overpowered by the magnificence of the rocks and scenery, Theo reported, "Every day of our stay seemed more glorious and wonderful."

On June 11, Theo and his father walked up Glacier Point and ascended the Sentinel Dome, a walk of nine miles and a rise of 4,125 feet above the valley. The same trip nine days later yielded an abundance of butterflies, as he reported to Mr. Edwards:

> On Monday, Papa, Mamma and I went up to Glacier Point, catching nothing of very great consequence; stayed there overnight and walked around next day to the head of the South Fork of the valley catching about 150 butterflies of considerable consequence, as the elevation was sufficient to offer several species entirely absent from the valley. Mamma distinguished herself by taking the rarest thing noted on the trip, a fine female *Argynnis epithore*.

After a visit to the Mariposa Big Trees Grove to view the majestic sequoias, they traveled north to Lake Tahoe, where they camped near Freel Peak, Jobs Peak, and Mount Tallac, close to the border between California and Nevada. They found a new and rare Alpine butterfly on August 1 in the proximity of Gilmore Lake, which Theo named *Chionobas ivallda* (Mead). In the Lake Tahoe/Donner Pass region, he found *Lycaena editha*, (Edith's Copper) which he named after Will Edwards' elder sister. In early July, he sent a barrel containing all his specimens so far collected, about 1,500 butterflies, to Mr. Edwards and promised to visit him at Coalburg in a couple of months to review the catches.

Still full of adventure and in no particular hurry to return home, they had previously bought railroad tickets from San Francisco to New York, "allowing for indefinite stop-overs anywhere." A leisurely perambulation followed in August involving further collecting, taking in Virginia City and a visit to the silver mines, Cottonwood Canyon, and Salt Lake City, and then the Green River Basin in Wyoming to examine the oil shale deposits containing impressions of leaves, insects, and fish.

In early September, they visited entomologists in Davenport and the Chicago area, before going to Coalburg to review the butterflies that he had collected and sent to Mr. Edwards. On September 23, his parents left Coalburg, taking the C&O to White Sulphur where they connected with the sleeper to Washington, thence to New York. Theo's father added up the expenses, ignoring food, for the three of them on their 188-day trip as; steamer $433.70, rail $604.07, horses $376.33, and boarding $1,079.42—approximately $3,000 in total. He wrote to his son, "I think we got the largest amount of fun out of the money possible which is precisely what money is good for." The account finished with Theo at Coalburg with $51.70 in his pocket, a through ticket back to New York, and the emotional squabble with the Delta Upsilon fraternity a distant memory.

CHAPTER 12

Alpha Delta Phi Brotherhood, 1878–1881

While Theo was away in California with his parents, Alpha Delta Phi had acted on building a Chapter House. His initiation into the fraternity had coincided with an internal debate about this possibility in early 1877 when most of the fellows were against the idea. But with the help of another brother, Theo had taken up the cause with enthusiasm. Cash was the issue, and Theo had turned to his parents for help. His mother had agreed to provide a $500 loan, but with strings attached of a religious nature. The deal she proposed to ensure his spiritual welfare was for him to attend an evangelical church service every Sunday for a year, and report back through letters with a description of the sermon. His father wrote on the envelope so Mary couldn't see it, "Don't do it, Ted. Charge $10 a time and stop when you get sick of it." Nevertheless, Theo agreed to the bargain and received the $500, even though he was still strongly at odds with his mother's view of organized religion.

With Theo's help, and letter writing to alumni and other potential donors, the society raised the necessary $12,000 in early 1878 to build the house, and had drafted plans by the time Theo left for California in the early spring. Ground was broken in June, and the house, on Buffalo Street halfway up East Hill, was completed in time for the start of the 1878 fall school year. Alpha Delta Phi became the first fraternity at Cornell to build its own chapter house—the two-story, brick house providing housing for sixteen brothers at a rent of $2.50 per week.

12.1: The Alpha Delta Phi Chapter House on Buffalo Street, and Theo in 1879, aged 27, proudly wearing his star & crescent pin.

On Theo's return to Cornell in October 1878, he was keen to be reunited with his fraternity brothers and to see the house. He was warmly welcomed, and praise heaped on him for the enthusiasm and energy he had expended in bringing the building to fruition. The attention he received, on top of the pleasant and relaxing experiences of the trip to California did the trick—Dudley and the many unpleasant episodes of the previous years were forgotten. His confidence returned, and he was back to his old self as a popular and active fraternity member, but now a considerably more mature one.

In late 1878, Theo picked up the post-graduate work that he had initially agreed to under Professor Prentiss, who told him he could make out his own course of subjects that were of interest. He chose about fifteen hours a week of coursework on an eclectic mix of subjects, veterinary science, modern history, organic chemistry, and English literacy, and twenty hours a week of laboratory classes in entomology and physiology.

12.2: *Theo sits on the floor, extreme right, in the 1878 photograph of the Cornell Alpha Delta Phi brothers. They are from left to right, with graduation year; back row: Catlin '82, Taylor '81, House '81, Fox '80, Gifford '79, Dr. Jones '75, Howland '79, Morris '80, Skinner '81, Washburn '83; middle row: Stambaugh '81, Smith '83, Edwards '79, Parmelee '81, Manniere '80, Howard '83; sitting on floor: Luckey '82, Booth, Q '81, Cushing '82, Saunders '81, Siras, W '82, Mead '77, and Jip the dog.*

Theo made many close friends in the fraternity. Notable were Edward Cole Howland (Howly), who graduated with a Bachelor degree in Literature and became a newspaper editor and writer for *The Washington Post* and *The Washington Herald*, and Henry Platt Cushing (Cush), who studied geology, graduated in 1882, and became Professor of Geology at Case Western Reserve University. His circle of friends also included Edward Mandell House, another literature student, whose party piece at the fraternity was to string together as many cuss words as possible into sentences, and energetically deliver them to a person in such a way that it sounded like he was pronouncing a benediction. He later rose to fame as the personal adviser to President Woodrow Wilson—Colonel House—who called him "my second personality." Harry Adams Robie was another close friend, who studied mechanical engineering and then tried his hand for more than ten years as a citrus grower in Mount Dora, Florida, where he was a neighbor and family friend of the Meads.

12.3: *Some of Theo's closest Alpha Delta Phi friends, left to right; Ed Howland (Howly), Henry Cushing (Cush), and Ed House.*

Theo's organizational abilities, attention to detail, and understanding of financial transactions and balance sheets were invaluable to the fraternity. Soon he became one of their leaders, taking on the role of treasurer and recorder and producing the first annual report for the Alpha Delta Phi Society in October 1879. There was always plenty for him to do. He organized the production of the all-important "Star & Crescent" fraternity pins that were essential ornamentation for identification and display, and started to collect data and material from all the Alpha Delta Phi chapters for the semi-centennial fraternity catalog, published in 1882.

Theo's best friend, Will Edwards, who graduated in the summer of 1879 with a Bachelor's Degree in Science & Letters, was always a practical advisor to Theo. He worried about how his friend was going to make a living in the future, and urged him to consider going to Harvard to do post-graduate work in natural history with Dr. Hermann Hagen, considering it a huge waste of talent not to do so. Theo explored this possibility on one of his fraternity tours, visiting the Harvard Chapter that included Theodore Roosevelt among its graduate brothers. He dined at Cambridge with Dr. Hagen, who showed him the biological collections that had been his life's work, introduced him to other Harvard professors, and spoke highly of Theo's pioneering work. But even with all this positive attention, Theo remained uncertain about further post-graduate work either at Cornell or Harvard, and instead was persuaded by his father to enroll for a year at the Columbia Law School to study law, following the paternal advice, "You must know law to keep out of it."

12.4: *Theo with an unknown dog in an image probably taken about 1880.*

In May 1880, Theo joined 125 other Brothers at the Thirty-third Alpha Delta Phi Convention at Rochester, New York. He wrote to his parents that he was enjoying himself but "had little time to sleep in the last few days," what with attending a "swell reception" and spending time getting to know all those Brothers he didn't already know.

Towards the end of 1880, his father found him a temporary position as a stockbroker at 115 Broadway in New York City, but this turned out to be of little interest and he returned to Ithaca. In January 1881, stained glass windows for the Chapter House ordered from Philadelphia arrived, and he was busy helping to set these in place. As Theo relaxed in the house, he reflected with quiet satisfaction that of all the fraternity chapter houses he'd seen over the last few years, the Cornell one was by far the best, and much of what had been achieved was a result of his efforts and energy. His membership of Alpha Delta Phi would be one he cherished for the rest of his life. In his view, its lessons in loving brotherhood were more valuable than all the textbooks of all the Universities. His beloved fraternity was thriving, but all his close Brothers had left to find their way in the world. He would stay in contact by letter with many of them and whenever possible visit them on his travels. But the big question was where was he going, and what was he going to do for the rest of his life?

Theo had matured considerably over his time at Cornell. In his formative years, he was easily upset and discouraged by criticism, and sought constant approval and "a few kind words" to reassure him that he was appreciated and had value. By the time he left Cornell, he had learned much about himself and developed a greater understanding of other people and their needs. He had matured, and elements of his character that were the product of a privileged but spoiled upbringing were beginning to fade. Will Edwards had said in 1873 that college life for four or five years at Cornell would take a whole lot of "calf" out of him—and so it proved to be.

Theo's interest in butterflies continued for the next few years, but it was clear to him that butterfly collecting and entomological activities were acceptable as a hobby, but could not fund his lifestyle. Although he had just graduated in the subject, Theo had no interest in being a civil engineer, and the brief exposure as a stockbroker in New York had convinced him that he wasn't cut out for deskwork either. His best childhood memories were of working in his father's garden at Fishkill, and since then he had kept the horticultural interests alive with his cactus plant collection and his first hybridization experiments in 1875 with the flowering *Abutilon*. After head scratching his way through the options, he concluded that probably his best bet would be to try to find a way to produce income from his passion for growing plants.

Butterfly collecting over the summer of 1880 had involved a tour of Newfoundland with his father. Collecting was a success, and many of the local Short-tailed Swallowtails (*Papilio brevicauda*) were taken in caterpillar form, and later hatched in New York to give an interesting series of variations. However, the trip was marred by the black flies that were a terrible handicap to outdoor hunting. As Theo later reported, "They went through fine mesh tarlatan like weasels through a rail fence." Each day soon after leaving the hotel, any exposed skin would be bloody and covered with irritating bites. They learned later that copious anointment with tar and lard would keep the black flies off. After three weeks of collecting they had had enough, and Theo suggested to his father that they should try a thousand miles south instead of north next summer, and see if it could be any worse.

As he pondered his options for the future, and with his mind becoming more and more focused on horticulture, a resolution to the fundamental question of how to make this interest income-producing emerged. Perhaps the answer did lay thousands of miles south—in Florida. There, the worse they might encounter would be redbugs, which by comparison to the black flies of Newfoundland were said to be only mildly annoying.

Part Four

A New Life in Florida

CHAPTER 13

Florida Land Purchases, 1881

By the second half of the nineteenth-century, the wealthy and curious inhabitants of the Northeastern states were increasingly reading newspaper articles portraying Florida as a dreamy, semi-tropical land of sunshine and relaxation. One of the earliest books promoting Florida tourism was Harriet Beecher Stowe's 1873 edition, *Palmetto-Leaves*. Subsequent serialization of chapters from the book and other essays appeared in the widely read *Christian Union* newspaper, which claimed 81,000 subscribers, many of them affluent New Yorkers. Writers such as Stowe described the luxuriant and fertile countryside, bursting with fragrant orange blossoms and other flowering trees, the delights of a perfect and healthy climate with spring-like weather, and the abundance of nearby lakes offering splendid fishing.

By the middle of the 1870s, the promotion was in full swing, and Florida started to experience the first of many development booms. Using brochures and newspaper advertisements, agents targeted the large northern and midwestern cities, selling the opportunity of buying Florida land to potential investors. They stressed the ease with which citrus could be grown and shipped to the lucrative markets further north, now the railroads had arrived. Readers were impressed to learn that a typical grove owner could pay for his land, recoup all his costs and start to be in profit in only three years from the time that his trees first bore fruit. Wrote one blurb, "It (citrus) is drawing men and capital from far and wide, and providing a source of industry and wealth for all who have patience and capital to cultivate it. The orange is surer than gold and more productive than grain or cotton."

> **It is Florida.**
>
> ★★★
>
> Knowest thou the land where the lemon trees bloom,
> Where the gold orange glows in the green thicket's gloom;
> Where the winds ever soft from the blue heavens blow,
> Where the stately magnolia and dark myrtle grow;
> Where beams the fair sun, where the mocking birds sing,
> And all the year offers the blessings of spring.

13.1: This verse appeared in the 1895 guidebook A Florida Guide for Tourists and Settlers.

A leading newspaper promoting Florida was *The Florida New-Yorker*, published in Orange County at Eustis. It started publication in 1876 and had the by-line, "Devoted, earnestly, to the good cause of Southern immigration." Another Eustis-based publication carried the editorial jingle, "No stormy winter enters here; 'tis joyous spring through all the year!" Adding to these and much other promotional imagery, such as the railroad and hotel pamphlets, were several Florida guidebooks aimed directly at tourists, invalids, and settlers.

Once the South Florida Railroad reached Orlando in October 1880, the boom accelerated, transforming the Florida peninsula and triggering a fifteen-year period of rapid growth and prosperity. Oranges became an accepted part of the winter holiday festivities for most Americans, and the railroads could deliver them to the northern cities within thirty-six hours of picking. Many of the oranges were grown in the Indian River locality, or within a very productive area of Central Florida between Mount Dora, Eustis, and Tavares, promoted as "The Golden Triangle." Here settlers developed groves in fertile soil by the shores of the rivers and many lakes, accessible to water for transportation. Eustis became known as "the orange capital of the world," and so it was natural that when Theo and his father arrived in Florida in July 1881, they came to this area in search of land and horticultural opportunities.

The man most often involved in selling this area of Central Florida to investors and settlers was John Macdonald, originally from Canada and the United States land office in Jacksonville. He advised potential investors on land buying opportunities, helped newcomers choose a homestead site, officially recorded their claims, and saw himself "at the pulse of emigration." He sold land principally in the area of Lake Eustis, where he helped establish a small lakeshore community, and by the mid-1880s he had acquired an impressive list of contacts and wealthy clients. After writing a plain-talking book about buying land in Florida and what investors should look for, he became so successful that frequently investors would leave the decision about where to buy entirely up to him.

Theo's father had a large portfolio of stock and property as his asset base that provided a regular income, but he was always on the lookout for further investment opportunities. He was aware of Macdonald's name so once citrus horticulture in Florida emerged as his son's intended vocation, Macdonald became his chosen advisor on where and what to purchase.

Theo and his parents arrived in Central Florida in late June 1881. They took the St. Johns River Steamer *Frederick de Barry* from Jacksonville to Enterprise, via Volusia and Sanford, where they disembarked to look at citrus grove property. On a side-trip to Orlando, they saw a fifty-acre grove of 270 orange trees with front and rear lake access priced at $15,000, which they considered too expensive. They then made their way to Lake Eustis with the intention of seeking out John MacDonald.

Their first impressions of accessibility to the region were good. Since 1880 the area had been connected to the St. Johns River transportation system by the St. Johns & Lake Eustis Railway Company, running track north to the village of Astor twenty-seven miles away. They were also enchanted with the summer climate, the local geography of the area, and lakes that were as clean as if supplied with pure spring water. Theo's father enjoyed regular swimming in the crystal-clear waters and wrote in his diary, "I think we would flourish here and hold on to life with all the tenacity of alligators." The only disadvantage he could find was the occasional worrisome fly swarms, but he had a solution, "Kerosene on eyelids, very good against flies," he wrote.

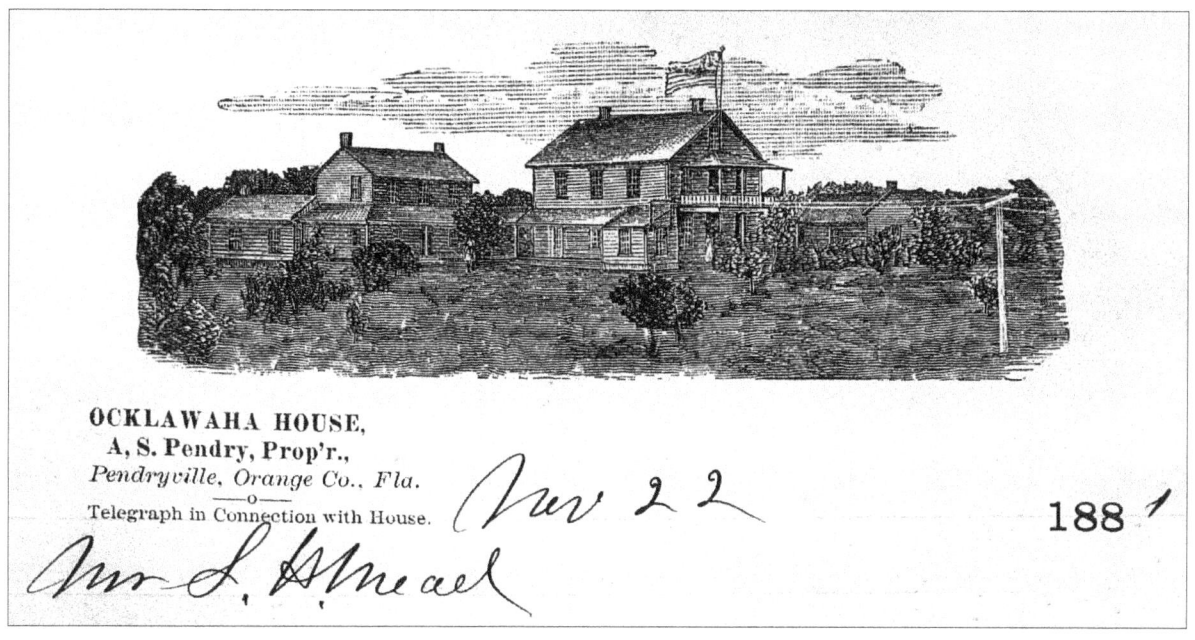

13.2: In the early 1880s, visitors to the Lake Eustis settlement, then known as Pendryville, stayed at the Ocklawaha House.

They stayed at the Ocklawaha House, the only tourist hotel in Pendryville,* and sought out MacDonald. He told them that although some undeveloped government land could still be found at $2 per acre, good agricultural land or land close to lakeshores and rich hummock areas was considerably more expensive, but if

---

*Pendryville subsequently became Lake Eustis, then in 1883 the "Lake" was dropped and the town just became Eustis.

they wanted he had both types to show them. A plan evolved to buy two parcels of land; the first as a speculative investment that would be left undeveloped; the second as a suitable place for Theo to begin his new life as a citrus grower and horticulturist. Income would come from citrus and the growing of other cash crops such as pineapples, leaving time for his experiments in semi-tropical plant growing and hybridization.

A mile or so from Eustis, MacDonald showed them a house, some outbuildings, adjacent land and a small orange grove. The house site, on an eighty-foot-high bluff overlooking miles of navigable Lake Eustis water, sloped steeply down to the lake and commanded one of the breeziest locations in the vicinity. Steamers passed by daily on their way to and from Lake Griffin and its environs, some sailing directly into the Ocklawaha River to Palatka and then by the St. Johns to Jacksonville. The beautiful view of the lake and the treetops far away along the shoreline acted on their senses. His father asked the price and in twenty minutes agreed to buy at $5,500. With the land deeded to him, Theo became a fully fledged orange grower. The property, a little over ninety acres, consisted of around fifteen acres of orange growing terrain close to the lake with adjoining high pineland to the east.

Later that month, approximately five miles east of the center of Eustis, they were shown land at $2 an acre. For $1,600, Theo's father bought an additional 800 acres as an investment, consisting of section 16 (640 acres) plus the southwest quarter of section 15 (160 acres) covering Lake of the Woods and part of Lake Mary. The terrain, mainly rolling pine woodland, was eventually leased for turpentine production.

Work started immediately on the Eustis Bluff site. The house, christened *Grove Cottage* by Theo, was habitable but lacked the convenience of running water. He decided to have a well sunk at the rear of the property supplying water to an elevated tank driven by windmill power. Mr. Pendry, owner of the local hotel, was contracted to sink the well. Theo also engaged help to start plowing land to expand the grove areas, and bought and planted out many orange trees together with peach, fig, pecan and banana plants. They had decided earlier that a useful cash crop before all the orange trees became fully fruit bearing would be pineapples, so a barrel containing a thousand slips was purchased. They set out individual plants between the rows of orange trees to benefit from whatever fertilizer the trees failed to take up.

13.3: In Florida, the ideal location for many settlers was an orange grove with a river or lake view.

The house was only about one and one-half miles from the downtown stores and post office of Eustis. Theo concluded that if all he wanted were his mail and a few light items, it would be faster to reach downtown by the waters of the lake, so he bought a boat. At the shore of Lake Eustis he had a boathouse built, reached by a sloping path down the bluff from *Grove Cottage*. Theo had already thought up a name for his venture,

Crescent Grove Experimental Farm, and had stationery produced with that as a letterhead.

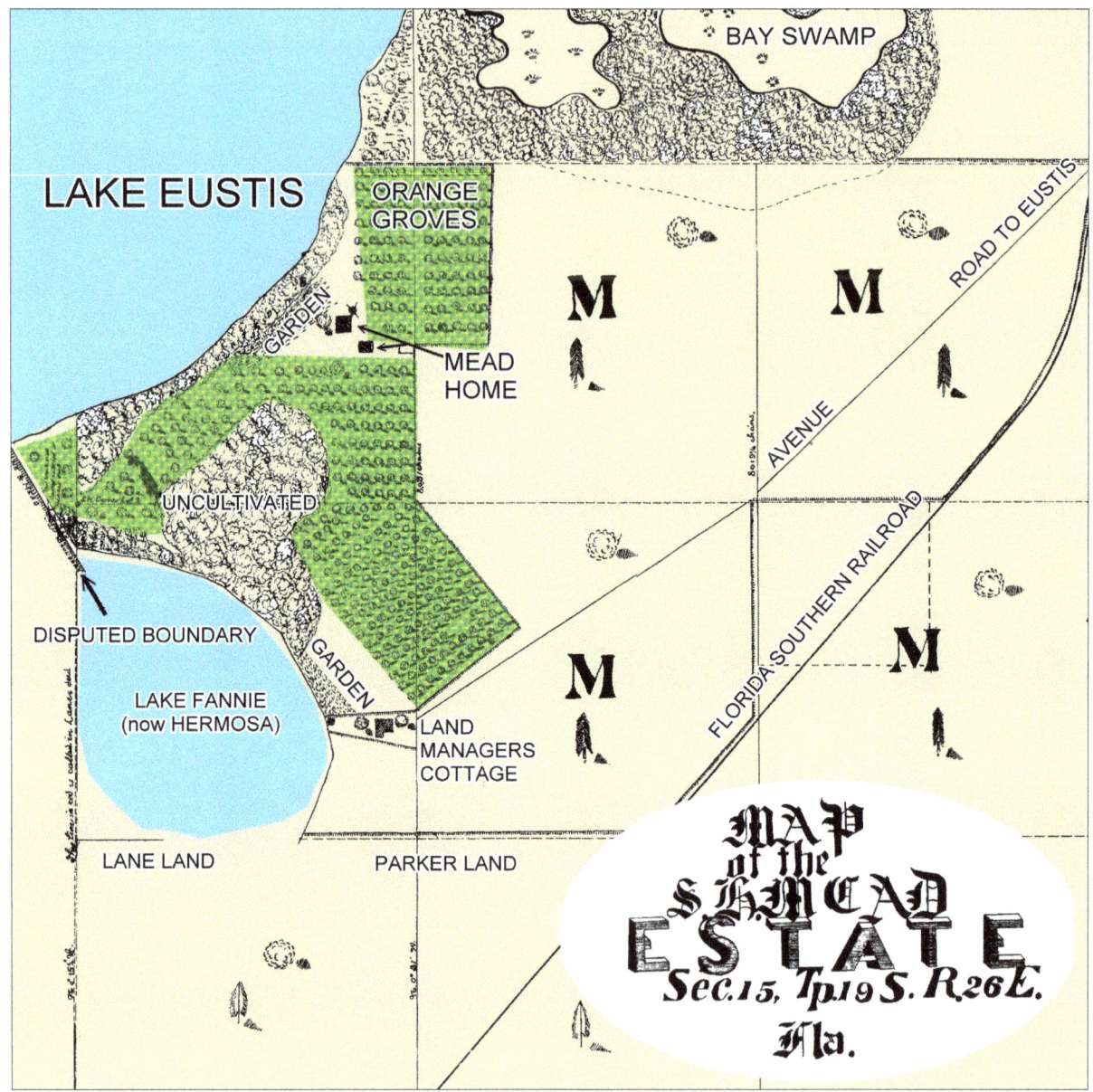

13.4: The 1890 plat of the Mead estate at Eustis Bluff, showing the orange groves, and position of buildings and cultivated lands by the lake. Today, residential development of the Springwood Landing subdivision in the top half of the plat has destroyed any trace of the groves and Mead dwellings, which were probably originally located on or close to Palmetto Road. The lower half of the estate remains undeveloped.

In the early part of the following year, Theo's father instructed MacDonald to purchase the adjoining southern two quarter-sections and the area close to the lake for $7,000, bringing the total acreage owned at Lake Eustis by mid-1882 to approximately 185 acres, and the investment to $12,500.

Theo returned to Philadelphia in November 1881 to finish off the Alpha Delta Phi semi-centennial catalog and tidy up loose ends. As 1881 drew to a close, he was enthusiastic about his future life in Florida and itching to get started, but first there was a crucial personal matter to be attended to.

CHAPTER 14

Marriage to Edith Edwards, 1882

Theo was approaching thirty years of age and dearly wanted children of his own. He logically concluded it was time he got married, but the problem was he did not know many eligible young women. Theo was a late bloomer in terms of his interest in the opposite sex, but over the years on his many visits to Coalburg, he had evolved an affinity and friendship with Edwards' two daughters, Edith and Anne. Both were pretty and charming, particularly the eldest, Edith, with her long, jet-black hair, whom he had first met in the summer of 1869 when they were both seventeen. With frequent visits, the relationship with all of the members of the Edwards family had grown, and he was always at ease and comfortable in their company. Theo and Edith discovered they had much in common; she was well read and mature and very much a match for his intellect.

Edith was born in London in 1852 when her father was there for a year promoting the Kanawha coalfields to investors. On August 8, 1853, shortly before the Edwards family returned to the USA, she was baptized Edith Katherine Antill in St George's Church, Bloomsbury, a stone's throw from the British Museum. The family sailed back to America from Liverpool on the steamship *Atlantic*, arriving in New York on the morning of September 4, 1853.

They returned to the family home in Hunter, New York, where Edwards fathered two further children—a son, William Seymour Edwards (Will) in 1856 who would play a significant role in Theo's early life—and a second daughter, Anne Scott Edwards in 1858. Edith was educated at Miss Porter's finishing school in Farmington, Connecticut, where among other things she studied literature, learned to play the piano, and ride a horse. In her twenties, she worked for a number of years as a nurse at the Homeopathic Hospital in Cumberland Street, Brooklyn. There she was a tireless worker providing humane and sensitive care to the sick and needy, which she believed was essential to promote healing. As an accomplished piano player, she was also in demand at the local church near Coalburg playing the organ, which Theo always said was partly his since he owned "one share of stock (one-tenth)" from his $5 contribution to buy it in the first place.

14.1: Edith Edwards in her teenage years and as a young woman in her twenties.

She also helped her father by drawing and coloring many of the chrysalis and caterpillars subsequently used as illustrations in his various publications on butterflies, although her younger sister proved to be a more skilled artist. The Edwards' house and outbuildings were completely given over to the life-cycle of the hundreds of butterfly species that formed his collection. Some areas were filled with masses of cuttings of the food plants on which caterpillars were reared—so much so that Edith and Anne had their own pet nickname for the house, *Raupen*, German for caterpillar.

As an honorary member of the Edwards family, Theo needed little in the way of an invitation to spend the Christmas of 1881 with them. He wrote to Will expressing the desire and received the enthusiastic response, "We are all looking forward to sharing the Christmas pig with the Florida planter. Come along the sooner the better!"

14.2: Bellefleur, *the Edwards' home in Coalburg, West Virginia, as it might have looked at Christmas 1881 when Theo proposed to Edith.*

Theo was a true gentleman, so would have approached the subject of marriage to Edith with her father first, who held Theo in high regard. He was delighted to have a fellow lepidopterist as a potential son-in-law, and possibly a little bit relieved that he had finally got round to asking her after around twelve years of acquaintance. Edith's acceptance spread happiness throughout the family, and the house was full of special joy over the already festive season.

For the last almost thirty years, Theo had been wholly funded by the considerable family wealth of cash, stocks and the asset base of several prime New York properties, representing their share of the Mead and Luqueer family fortunes. His parents had willingly supported him through his periods of education, both at home and abroad, his travel expeditions to collect butterflies and attend to Alpha Delta Phi fraternity business, as well as his day-to-day living expenses. Now with his forthcoming marriage to Edith, there was extra pressure on him to produce an income of his own rather than continue to rely on his parent's generosity.

Over the Christmas period at Coalburg, the Edwards family had overwhelmingly supported the view that citrus growing in Florida was the answer to financial independence. They pointed to the direct experience of Edith's Aunt Mary, who had married Dr. Henry Foster of Clifton Springs in 1872. He had bought orange-growing land at Lake Charm, near Oviedo in Central Florida and although the couple were only there for a couple of months over the winter, and paid hired hands to manage the groves for the rest of the year, the land had proved very productive and generated significant income. By the time Theo left Coalburg on the evening of January 8, 1882, engaged to Edith, he was confident he was on the right track with the Eustis property.

They had decided to get married in early June, followed by a honeymoon in England. The many letters that passed between the two of them before this provided an opportunity to discuss the wedding and honeymoon arrangements, the finer points of their relationship and how they were going to live, as well as cement the love they felt for each other.

Theo had always been wary of ferociously religious people, women like his mother intent on passing their pious beliefs directly down to their children, as Mary had attempted with him. He did not want this to happen to a child of his, so now that marriage was imminent a major issue was Edith's precise religious orientation. He knew she was a Christian certainly, but since she was descended from the great revivalist of the eighteenth-century, Jonathan Edwards, Theo wanted to know whether she classified herself as a revivalist, undoubtedly worried she might, in religious terms, turn out to be a carbon copy of his mother. Edith handled this question with great tact, replying, "You won't have to fear having chosen a 'female revivalist,' for as you know I don't approve of that sort of thing. I believe in complete freedom of conscience, and shall never try to make you, dear boy, try to believe as I do where you can't. At the same time of course it will be a great happiness whenever you can and do."

Theo had found a real treasure—a bright and charming woman with an inquiring and intelligent mind, who was kind and considerate and would be a source of great strength and support to him in his new horticultural endeavors.

14.3: A treasured wedding gift in June 1882 was a Greek bronze vase given by the Alpha Delta Phi brothers at Cornell, and sent by Theo's friend Howard Gifford.

102   Orchids and Butterflies

Theo arrived at Coalburg in late May a few days before the wedding. After the greetings were over, Will brought in a mysterious package and left it for Edith and Theo to open. The sender was his Cornell Alpha Delta Phi brother, Howard Gifford, and the package contained a reproduction classical bronze vase with the inscription, "Presented to Edith Antill Mead and Theodore Luqueer Mead from '78, '79, '80, '81 in Alpha Delta Phi, Cornell University."

At the wedding ceremony at *Bellefleur* on Thursday, June 1, 1882, Mr. John Lea officiated and Webster Smith, soon to be Anne Edwards' husband later that year, signed the marriage register. Although no photographic records of the wedding have survived, a letter from Edith's mother Kate to her sister Sarah provides a vivid description of the ceremony itself:

> By eleven o'clock the guests began to arrive, and at half past Edith was dressed and ready. You would have approved of her simple costume and have agreed with us that it was much more suitable for the time and place than a more gorgeous one would have been. The pure white nun's veiling trimmed with many rows of lace and with drapery of white watered ribbon at the back was very pretty. The veil, a large square of tulle, fell a little over her face and was caught on one side of the head by a cluster of natural white roses. Another bunch at the left side of the neck among the laces and a third larger upon the skirts holding the veils back a little. Black stockings and slippers and long white gloves made up the ensemble. The only ornaments were a beautiful silver comb & earrings sent her by Cousin Hannah Forbes. Every one said she looked lovely only very pale. But she was perfectly calm and composed.

Mary Mead was described as "looking uncommonly well in black satin and lace, some beautiful diamonds at her throat," and both parents expressed great happiness in having, at last, "a beloved daughter." Taking the C&O railroad east, the newlyweds spent their wedding night at the hotel at White Sulphur Springs, West Virginia, where they stayed for three or four nights before reaching New York on June 5.

14.4: *Theo's father, sporting magnificent mutton-chop whiskers, and his mother, who usually dressed in black, were overjoyed at the marriage of their son to Edith.*

Mr. Theodore Luqueer Mead,

Miss Edith Antill Edwards,

Married,

Thursday, June first, 1882.

Coalburgh, West Virginia.

At Home
after October first,
Lake Eustis, Florida.

14.5: *The Mead's official wedding notice.*

CHAPTER 15

Honeymoon in England, 1882

Theo and Edith left New York by steamer to England on June 10 and arrived at Liverpool eleven days later on June 21. The crossing was slow but uneventful, although his parents were anxious because an unusually large number of icebergs had been spotted and reported in the New York newspapers. His father wrote, "We have accounts from incoming ships of large numbers of icebergs seen sixty at a time and ice to the horizon all around; fogs shut down and enveloped them until their arrival at New York. You landed on Wednesday evening 21st, 11 days and some hours from New York and probably escaped these annoyances. Hope you had a good time."

From Liverpool, they made their way by train via Manchester to Buxton in Derbyshire, a fashionable resort and spa town where they intended to spend several days and take the waters. By the end of June, they were in London, initially taking rooms at the St Pancras Hotel on Euston Road. From there they visited the colleges of Cambridge and the London Zoo, which prompted Theo to report, "Yesterday afternoon we went to the Zoo and were entertained especially by the fine collection of parrots—they made a deafening screeching but most were tame and seemed to enjoy having their heads scratched. Even the big macaws on perches around the room were as tame as could be and friendly." In early July, they found accommodation on the south side of Bernard Street, near Russell Square and the British Museum.

Letters between Theo and his parents were infrequent but did not stop his mother worrying about her thirty-year-old son and the temptations of a big city, which might result in him straying from the path of righteousness. Part of one letter read:

> I trust there is no alcohol in it from London stout, porter, ale or any foe of that kind that vanquishes thousands. Yesterday afternoon I attended a temperance meeting in a tent and listened to the simple tale of several reformed men of their degradation and fall, struggles with appetite, and their deliverance. But it is said that it is impossible to eradicate entirely the taste and love for it. How foolish and hurtful to acquire the habit.

Visits were made to Covent Garden, where Theo was astonished to see imported pineapples at 6 guineas each (roughly $50), to the British Museum to see the Elgin Marbles and the Egyptian remains, to the Tower of London to see the Crown Jewels, and to Madame Tussauds to view effigies of famous people, where they were well entertained. "Edith says she will dream of the Chamber of Horrors tonight," wrote Theo to his parents.

While Edith rested, Theo visited Kew Gardens on two separate occasions, complaining each time about the problems of getting there by train. However, once there he was impressed with the cactus collection and wonderful Victorian-inspired Palm House, reporting, "The great Palm House is a complete picture of what the tropics ought to be and generally are not, enormous palms and screw pines of amazing size—branching and with each head twenty feet in diameter, every leaf ten feet long and more humble plants flowering or ornamental or curious filling up all available space."

15.1: Inside the Great Palm House at Kew.

At the end of the first week of August, after almost five weeks in London, they were becoming a little tired of the city and yearned for some fresh air. They took a train to Chester and prepared to return home, buying a steamer trunk for all they had purchased or collected.

In mid-August, they departed from Liverpool bound for New York on the White Star Line *SS Germanic*, arriving at 5:00 a.m. on the morning of August 27 after an uneventful crossing. After a short stay with his parents in New York, Theo and Edith traveled to Coalburg to attend the wedding of Edith's younger sister, Anne, set for October 5. She was to marry Webster Dryden Martelle Smith, originally from Maine and an 1878 graduate of Dartmouth College, who worked as a traveling agent for the Dixie Dynamite Company of Chattanooga, selling black powder to the coal mine operators in the area.

15.2: Webster Smith and Anne Edwards some time in the early 1880s.

A few days of rest and recreation followed with long walks up Paint and Cabin Creeks. Theo was impressed by the progress of mining development in the area and astonished to find a broad gauge railroad running for several miles up Cabin Creek to reach the coal mines. With the fall colors just starting, Theo wrote to his parents, "The scenery down the river was much more beautiful than any seen in England."

In late October, Theo and Edith traveled by train to Richmond where they were due to meet up with his parents coming down from New York. Together they made their way to Lake Eustis, where he was to begin his new life as a citrus and tropical plant grower.

CHAPTER 16

The Eustis Years, 1883–1886

At Eustis Bluff, Theo, his father, and the hired hand Nixon, set to work hoeing and heavily fertilizing the existing orange groves consisting of almost 600 trees of varying maturity. They created two new groves close to *Grove Cottage* by planting over 400 sour-orange rootstock seedlings. Theo's parents returned to New York in mid-May leaving them with Harry Norton, a local orange growing specialist, who budded the seedlings with various varieties, such as Mediterranean Sweet, Majorca, Jaffa, Satsuma and Maltese Blood. Theo was full of enthusiasm for the progress made. He had six groves with over one thousand trees—Big Grove (320), Summit Grove (91), Log Cabin Grove (70), Lake Shore Grove (254), Crescent Grove (97) and Cottage Grove (210)—and by early summer of 1883, all the trees were growing well.

He positioned many ornamental plants along the walk from the house to the lake and added guava and loquat trees to areas already planted with peach, fig, pecan, and bananas. At the north corner of the land, he located his cactus and succulent collection, including a plantation of various species of agaves and yuccas. Directly outside the house, he planted climbing vines that would eventually entwine and cover large areas of the structure.

By mid-summer 1883, the Meads were itching to return north to hear the family news and the sound of familiar voices, so in late July they traveled north to Coalburg. In September, they went via Theo's parents in New York to the Edwards family home in the Catskills, staying with Edith's grandparents. Edith returned to Coalburg while Theo went on to Ithaca to see his fraternity friends in Alpha Delta Phi. When it was time to leave, Edith's mother accompanied them, and they all arrived back at Eustis at the end of October. Mrs. Edwards' first impressions of Florida were very positive—she was delighted with the quality of the air, the fragrance of the orange groves, and the beautiful view across the lake from the bluff.

It had been a hot, dry summer while they had been away, and this had led to several plant losses, including most of the pineapple slips. This forced Theo to think again about growing pineapples, and he consulted with other growers. They told him that there were only small pockets of land in Florida where the soil was humus-rich and suitable for pineapple production. A further limitation was the pineapple's sensitivity to cold, which could easily check growth and cause them to ripen early, yielding only small fruit. But Theo was determined to succeed, spurred on by his memory of seeing them selling for $50 each in a Covent Garden market stall in London in 1882. His solution was a boarded pineapple pit twelve by six feet and three feet deep, in which he proposed to plant eighteen plants in a mixture of animal waste and good top soil, so that the temperature of the mix would remain roughly constant, warmed by the heat of decomposition of the manure.

The winter of 1883–1884 was a cold one, with freezing temperatures hitting large parts of the region. Eustis Bluff was spared so his oranges were much in demand, and he was able to sell them for 2 cents each by the dozen and $15 a thousand. His only problem was there was a good deal of stealing of fruit from trees located near the edge of his property, where he counted only around 200 to 250 fruit per tree. He sent several boxes of mixed fruit off to various relatives and wrote to his parents, "So far I haven't seen a single orange tree that has as good a color as our Log Cabin grove."

Fertilizing the trees continued into the summer with Theo and Nixon distributing 1,000 lb. on the Log Cabin grove—about 13 lb. to each large tree. They gave the large trees in the Lake Shore 6 lb. each and the smaller ones half that. In July, Theo counted 2,135 oranges on the 67 bearing trees plus approximately 5,000 lemons on 120 trees, with all trees growing well. His enthusiasm boiled over in March 1884 with a hypothetical calculation of his citrus business in the future when he might have 2,000 trees and generate $40,000 in one year from a harvest of 2,000,000 oranges selling at 2 cents each.

Theo was in an ebullient mood over his citrus success. He was thirty-two years old, weighed one hundred and forty pounds, and was physically fit and tanned from all the outdoor work in the sun. Before leaving to return to Coalburg the year before, Mrs. Edwards had told him she had never seen him looking so well; that he was the picture of robust health and should be exhibited as an advertisement for Florida.

16.1: Theo reclining on a live oak tree festooned with Spanish moss at his Eustis estate.

By comparison, Edith suffered from occasional debilitating headaches and from tiredness in periods of excessive heat and humidity. Life was physically hard and her condition was exacerbated by the irregularity of servant help. Servants came and went with the result that there were frequent extended periods of time when Edith had to do all the cleaning, cooking, and washing, contributing to her feeling of being worn out. They had trouble understanding why getting reliable servants from the local population was so difficult. Theo complained to his parents, "Even Colonel Lane's negroes refuse to serve us in any way although the family are away and they must have little or nothing to do. We get on very well but I hate to see Edie near the hot stove in this sultry weather and except for baking bread we shall probably rely on the kerosene stoves till we get help."

Inspired by his visits in previous years to the great palm houses of Europe, Theo started collecting and growing palms with the objective of finding which ones could be successfully grown in Florida. He ordered palm seeds from all over the world, bought small seedlings from growers, and built a unique collection of over eighty species, doing business on a regular basis with Haage & Schmidt, in Erfurt, Germany. On one occasion, he received one hundred seeds of a rare palm—*Erythea edulis*—mailed to him from a dealer in San Diego. Theo's passion for palms had him planting his favorite one, *Cocos flexuosa,* in front of the house, and *Cocos bonnetii,* another favorite, with its edible pindo fruit that Edith often made into jelly. Edith had never liked the "Crescent Grove" in the farm address, considering it rather common, so in light of his interest in palms, Theo changed it to the Ponemah Palmery, Box 15, Eustis, Florida.

Palm seed germination required patience and was a slow process, as he wrote to his parents, "Yesterday I found a Coquito palm coming up—seed planted thirteen months ago. It is a hardy Chilean species (*Jubaea spectabilis*) allied to Cocos but of rather slow growth. Still it is a species which I have long desired to possess." One of the slowest palms to come up was *Acrocomia totai*. He had seeds in sprouting pots for four years and three months before the first seedling shoot appeared, and after five years, a number of others germinated in the same lot of seed.

16.2: Two of Theo's favorite palms; the elegant Cocos flexuosa *and* Jubaea spectabilis, *the Chilean Wine Palm.*

By the summertime, his ornamental and other semi-tropical plants were doing well, aided by the wheelbarrow load of muck that he religiously dug into the soil at planting time. In front of one of the houses, he had rose bushes blooming and a little further away, several Eucalyptus plants destined to reach tree size—*calophylla, globulus compacta, cornute,* and one "with deliciously scented foliage exactly like lemon verbena"—*Eucalyptus citriodora.* Color was accentuated around the property with the Mexican creeper or coral vine (*Antigonon leptopus*) and several *Mandevilla urophylla,* with deep salmon blossoms and bright yellow throats.

His father was impressed with Theo's industry and his accurate record keeping, but the monthly farm expenses were generally more than $100, and more in those months when he needed to buy fertilizer. For his part, Theo was concerned that although they were not living in a wasteful manner, money paid out exceeded their meager income. To keep things afloat required cash from his father on a regular basis. The family still had plenty of money, but most of it was tied up in real estate and stocks, so his father frequently had to sell parcels of shares even when the timing was not advantageous.

In March 1884, an opportunity arose for Theo to raise some cash himself by selling his magnificent butterfly collection, still in New York at his parent's home. He had a visit from the Reverend Dr. W. J. Holland, who knew of Mead's reputation as a leading lepidopterist and had traveled to Florida primarily to see him and buy his collection. Holland was an avid butterfly enthusiast and being very wealthy, amassed other people's collecting efforts through the power of his checkbook, so eventually they struck a deal.

A typical day at Eustis Bluff began with Theo rising about six o'clock, lighting the kerosene stoves and putting the kettle on. Still in his nightshirt, he would pull on a pair of shoes, let the chickens out of their hen house into the run and feed them, then stroll around checking his ornamental plants near the house until Edith appeared downstairs. She had to remind him that if he wanted breakfast he needed to get dressed first. The first meal of the day was usually coffee and bread toasted on the griddle, liberally spread with homemade guava jelly. After breakfast, he devoted the next hour or so to answering the letters he'd received the day before and writing any new ones. Then he would see to the horse and start work with Nixon around the estate. There was generally work to do in the groves, but failing that, there was always cutting the lawn or tending to his ornamentals and plants in the vegetable and fruit gardens.

Lunch was a simple affair of perhaps bread and eggs, which they consumed in quantity, getting around a dozen a day from their flock of hens. They ate meat and fish generally only on the weekends. Once or twice a week in the afternoon, to coincide with the arrival of the mail, Theo and Edith took the wagon into

Eustis; on the other days either Nixon would collect the mail or Theo would take the boat if the weather was fine. The village of Eustis was growing; there was an excellent restaurant, a telegraph office, and a four-page weekly newspaper, *The Semi-Tropical*, published in a small building on a wharf at the end of Orange Avenue. Several sawmills had been established in the locality to supply lumber for building construction, and between the two main stores, Smiths and Cliffords, stood the big hotel, the Ocklawaha House. While in the village, Theo could cash any checks from his father at Smiths, buy food supplies and sell any surplus eggs before visiting the post office for the mail. Overall, there was a general feeling in the town that Eustis would grow to be an important urban center for Central Florida and prosper accordingly.

From time to time, Theo went out on Lake Eustis and fished for bass from his boat, and sometimes caught catfish from a baited line running out into the lake from the boathouse. Occasionally, in the heat of summer, and after several hours of hard physical work, he would go down the sloping path to the boathouse on the lake and surf bathe to cool down.

A further routine was the boiling of well water to make sufficient quantity of drinking water for daily consumption. During their time in Florida, all the family had occasional bouts of malaria. Theo kept up with medical matters and followed the accepted wisdom at the time that the illness was associated with drinking contaminated water containing malaria microbes. It would be the end of the century before the real villain—the mosquito—was identified as the carrier; until then combinations of contaminated water, malarious air, and swampy ground were variously held responsible for outbreaks throughout Florida.

Around 4:30 p.m., he took a break for half an hour or so and read the mail and any newspapers that had arrived that day, then in summer worked until around 7 p.m., watering and tending to his greenhouse plants. A bath and supper took him to 7:30. On Saturdays and Sundays, they finished their evening meal with the indulgent treat of ice cream. He told his parents, "Ice is 1¢ a pound or 75¢ a hundred, so we indulge in a quarter's worth once a week and have ice cream.

I made a freezer out of the little bloater keg and Edie uses the patent cylindrical egg beater and cream whipper to freeze the cream in."

After supper, there was the routine of reading aloud to Edith while she knitted or hand sewed until bedtime at 8:30. Books were read at a prodigious rate and their tastes were wide-ranging, covering classic novels, historical biographies, travel books, and religious tomes, as Theo described, "Edie and I are reading Haeckel's *India and Ceylon* and find it an interesting book of travels. Last evening I read Edie one of the *Broken Shaft* tales, greatly to her edification. The night before I read her *Dr. Jekyll and Mr. Hyde* and wouldn't stop till the tale was done so we didn't get to bed until nearly midnight."

The intellectual pastime of reading was rare in the area and they both missed the familiar cultured minds of their Northern friends. Theo made no bones about the way he felt about this, writing to his parents that except their immediate neighbors, the Parkers, "The other inhabitants of the region round about are almost exclusively Western farmers whose fields are much more cultivated than their minds." Some intellectual relief was at hand however with the arrival in 1884 in the nearby village of Mount Dora of an old Cornell Alpha Delta Phi brother, Harry Robie.

Theo had received a letter in early 1883 from Harry, who was then living and working as a mechanical engineer in the Navy Yard in Boston. Captivated by glowing accounts in the media of life in the 'sunny south', and the financial rewards of citrus growing, he wanted to know from Theo whether he should consider relocating to Florida, and if he did, where might be the best location.

Theo had replied positively and as a result, Harry bought 20 acres of orange grove in March 1884 in Mount Dora for $600, in the southeast corner of section 33, township 19, range 27 east. Harry married Alma Hodges in New York State in late 1885, and they made their home in Mount Dora at the *Colina* citrus grove. Through their friendship, the two men exchanged useful knowledge and information—Theo passing on his horticultural experiences and Harry, a mechanical engineer, his working knowledge of operating steam engines, pressure boilers, and well-pumping equipment. A warm friendship also developed between Alma and Edith, and they saw each other frequently even when the Meads had moved from Eustis. The Robies had two sons, the youngest born at Mount Dora, and they eventually became as precious to Theo as if they were his own.

16.3: Harry Robie, a Cornell Alpha Delta Phi brother, established the Colina Grove in Mount Dora in 1884.

CHAPTER 17

Dr. Henry Foster, 1885–1886

Theo and Edith had family in Central Florida. Thirty miles southeast of Eustis, near Oviedo, and on the shores of Lake Charm, was the winter home of the Fosters, Edith's aunt and uncle by marriage.

Henry Foster was born in 1821 in Norwich, Vermont and studied medicine, concentrating on homeopathy and hydrotherapy as part of his post-medical training. An intensely religious man, Foster believed in the efficacy of water cures and furthermore that he had a mandate from God to establish one based on Christian principles. This had brought him in 1849 to Clifton Springs, New York, which had a reputation for sulfurous waters similar to those found at White Sulphur Springs, West Virginia, one of the most popular resorts in the South. Convinced that this was where God had led him to start his work, Foster purchased the wild sulfur brook and marsh with ten acres of ground for $750, and began constructing a new building—The Water Cure building—that formally opened on September 13, 1850.

Foster practiced patient-centered care with minimal medication and maximum exposure to evangelical religion and the healing powers of spa water. Guests were educated about the importance of fresh air, good eating habits, exercise, and prayer. The consumption of tea and coffee was prohibited and meat used very sparingly. On a regular basis, in accord with Christian principles, the Sanitarium provided free or low-cost rest and recuperation for poorer people who could not afford the standard treatment rates.

17.1: Dr. Henry Foster and Edith's Aunt Mary occupied their winter residence on the shores of Lake Charm, near Oviedo.

Henry Foster had met his future wife, Mary Edwards, in 1871 when she came to the sanitarium as a companion to her invalid father, and Edith's grandfather, Mr. W. W. Edwards of Hunter, New York. On Wednesday, June 19, 1872, Henry Foster and Mary Edwards were married in a ceremony officiated by the Reverend Mr. Howard.

Foster believed in doing God's work intensively for ten months of the year and taking a prolonged annual vacation to rest and recover. Florida was a favorite place to relax over the winter months, spending time hunting, horseback riding, and fishing. In 1867, he bought a steam yacht, christened it *The Curlew* and took to the waters of the St. Johns River. *The Curlew* was a large steamer about 35 feet long, with an upper deck for shade and capable of carrying up to twenty people. On one occasion, sailing near Lake Jesup, he was asked to attend to the wife of Mr. Walter Gwynn, who owned much of the land extending about a mile south and close to a small lake, Lake Charm, and the Lake Jesup Community, later renamed Oviedo. Mrs. Gwynn was so ill that after treating her, he took her back with him to Clifton Springs when his vacation was over, caring for her for almost a year until she regained full health, all at no cost to Mr. Gwynn. A grateful Walter Gwynn wanted to give Dr. Foster some of his lands instead of payment. Foster refused but Gwynn insisted, so they agreed to a compromise involving deeding twenty-five acres to his new bride. Foster bought additional orange grove land from him, on which they built their winter home, *Rest Haven*. The grove land consisted of many muck beds—areas of decomposed leaf mold and other organic material on the sites of old lake beds and swampy ground—providing an inexpensive source of fertilizer.

Henry Foster became the Lake Charm community's chief benefactor, encouraging wealthy Northerners to over-winter there and actively promoting all aspects of the area's development. In addition, with the Mead, Edwards and Foster families connected through marriage, all family members would use the services of Clifton Springs Sanitarium and Dr. Foster when illness struck.

In March 1884, the Fosters invited Theo and Edie to *Rest Haven* and took them on a Lake Jesup excursion on his yacht. Once on the lake, they transferred to a small rowboat, trolled for fish, and soon caught four excellent black bass weighing from four to seven pounds. A little more steaming brought them to Gee Hammock where they picnicked on fried fish and other food they had brought. Theo examined the trees there and commented to his parents, "I think ten acres of the (worthless) "low hammock" between this and Lake Jesup is better than all the waters of Eustis."

Back at Eustis, Theo began to look seriously at the question of using muck as a fertilizer. Muck from beds on his land was free for the taking, but labor had to be expended in the digging, treating and transportation of it to the groves. Closeness was key—if the muck had to be carted over miles of rough Florida tracks, there would be far less difference in cost compared with commercially available fertilizer. Theo's calculations indicated a saving of around $125 per acre for his local muck, but he had only a limited supply and after a season or two, he had to purchase fertilizer again.

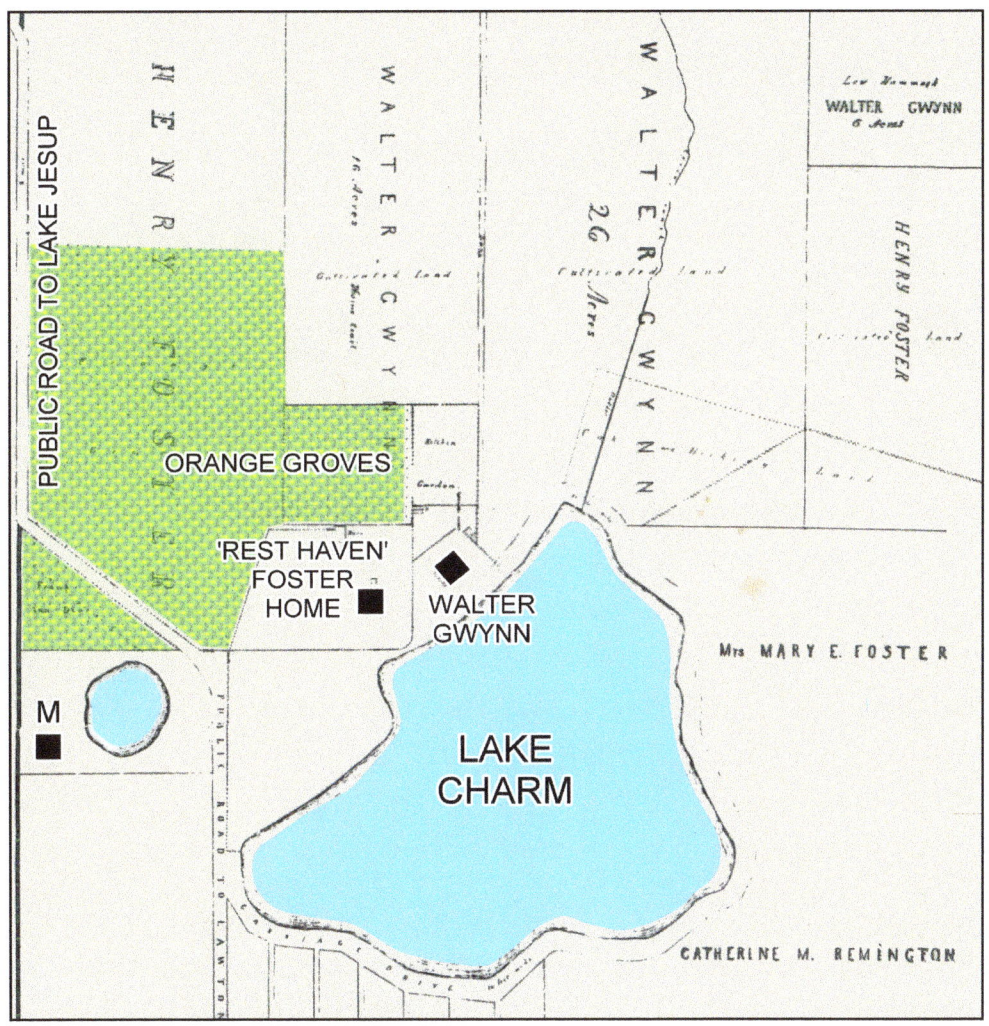

17.2: *Map of Lake Charm, redrawn from one circa 1870, showing the Foster and Gwynn homes and associated land. In 1882, Plot M, and areas to the west shown in more detail in 18.1, would be bought by the Meads and be the site of their new home.*

17.3: In 1884, Dr. Foster took Theo and Edie out on Lake Jesup in his steam yacht The Curlew. *This Florida period photograph captures something of the flavor of that day.*

Theo always knew that it might take him up to five years for his groves to become profitable, but on examining the annual farm account in June 1885, the gap between his expectations and reality was wider than he expected. In their first full year, they recovered one-fiftieth of their annual running expenses and now things had only improved slightly to around a twentieth. In response, he started to supplement their income with sales of his nursery plants and fruit and vegetable crops, although it was small potatoes in cash terms. He sold 2½ pecks of guavas for 40¢ and advertised in the *Florida Agriculturist* offering *Antholyza* bulbs for 10¢ each and young Surinam Cherry plants at six for a dollar.

His extensive collection of exotic palms was very much a mixed bag regarding their survival over the winter. Some had succumbed to the early winter cold, the decayed centers pulling out readily; others had escaped drastic damage and

122   Orchids and Butterflies

were candidates to add to the flora of Central Florida, such as *Cocos australis* and *campestris, Phoenix dactylifera* and *Canariensis, Chamoerops humilis* and *Sabal Blackburniana.*

The tomatoes and banana melons were producing plenty of fruit, but many were lost to worms boring in and consuming the interiors. The figs were doing little or nothing, although a few remained alive, and the lemons were rustier than ever, with numbers of good fruit at about one-third of last year's crop. The orange harvest was down too, and Theo guessed the number of oranges in the North Grove at about four thousand. In the spring of 1885, in anticipation of securing higher prices he had delayed picking all his oranges, leaving a thousand or more on the trees only to discover that the woodpeckers drilled into each one that didn't naturally fall from the trees. In November 1885, oranges by the box were quoted in New York at $1.75, which Theo ruefully reflected after deducting expenses (box 20¢, packing 25¢, freight 75¢), left but 55 cents for the grower. He concluded as far as oranges were concerned that "I doubt if they ever do more than return their cost of cultivation here."

Theo's visit to Dr. Foster's establishment at Lake Charm had created uncertainty in his mind over whether he had chosen the right location. The glowing reports of the amount of money Foster was making on citrus still rang in his ears compared to what he was expecting to earn that year. Although things had not gone badly, after eighteen months at Eustis, he was beginning to feel discouraged. Despite all his hard work, everything appeared to be flourishing on the Foster estate, without the outpouring of expensive fertilizer or the kind of horticultural knowledge he was providing at Eustis. They had chosen the Lake Eustis property on account of its frost-protected location, its splendid lake view from the bluff, and the immediate access to the lake. Also, the land prices there were such that they could afford a much larger estate than would have been possible around Lake Jesup. Now he wasn't so certain he'd made the right choice, and started to worry that they should have investigated more and talked to Foster about the advantages of the Lake Jesup region.

Hanging over him, and no doubt contributing to his overall discouraged mood, was the lack of progress in his desire to have children of his own. Edith, now in her early 30s, had not yet become pregnant after two years of marriage, whereas her younger sister Anne, married several months after her, had already given birth to her first born, Catherine Tappan Smith. The question of adoption had come up early in their relationship, and they had agreed that if they were still childless after three years, they would consider adoption, with Theo stipulating that it should be a boy of good parentage and Edith secretly wishing for a girl.

By early 1886, this doubt about location began to affect him, and his letters to his parents became increasingly despondent. He concluded they had made a serious mistake in choosing the Lake Eustis location for horticultural purposes. From extreme and sometimes naïve optimism about the financial rewards of citrus growing in the area, he cycled swiftly to the other extreme. He began to turn his exasperation against the region and its inhabitants, pulling no punches with his criticisms. He wrote to his parents, "The general tone of the 'Eustis Lake Region' I think, is an excellent average sample of the style of Eustis people and it is bucolically Philistine to the last degree, so much so that even I who am most tolerant of almost every form of crudity, get tired and disgusted with it."

His opinion of the soil at the Eustis Bluff location had turned equally sour, describing the "ill-adaptedness of the location to grow anything" and the "intrinsic wretchedness of all this soil," ending one letter with the heavily underlined sentence "The chief requisite to me is that things should _grow_." He analyzed the problem as being due to the combination of a western slope, exposure to the wind, and a largely sandy, humus-poor soil.

Theo was all for selling the estate if possible and moving to a more suitable location. But the initial land boom appeared to be over, and they were not certain they would even get their original stake back. His father, rather than criticizing his son and reminding him that he was a key part of the purchase decision in the first place, assumed full responsibility, declaring:

Certainly sell the whole plot if possible. You are the one to be the proper judge of what you would like. I made a great mistake in influencing your judgment in the purchase of the Herrick place. I looked at it only from my own standpoint as a place likely to advance in value. The other mistake was in pouring out our double-eagles on that kind of soil. The original purchase was not bad as a speculation.

Theo had the family go-ahead to reset his location, but also his fundamental business imperatives to make sure he was not solely dependent on citrus. To sustain his interest in horticulture, and knowing of his love for semi-tropical plants such as orchids, his mother had attended an orchid sale in New York without his knowledge and had successfully bid on a large mixed box of a hundred or so unknown orchid off-shoots, which she sent on to him. Theo was delighted with the haul and although he had a few orchids at that time, this shipment in November 1885 appears to be the first time he considered the possibility of orchid growing in a serious way.

Despite Theo's growing sense of disillusion about horticulture at Eustis, his father loved the bluff and the scenic beauty of the location, calling it "an ideal occupation or existence. I do not see any so altogether admissible and satisfactory. It is healthy, prospectively profitable, intellectual, and aesthetic. The water or lake part of it was the daydream of my boyhood. In all visions of possible bliss the foreground was filled by a smooth piece of water."

Still his parents were worried—they desperately wanted Theo and Edith to be happy, settled, and above all accessible. They were concerned that Theo had mentioned some hair-brained scheme about a possible move to Cuba to grow orchids. His father expressed his views forcefully, "I don't care a tinker's damn (that's not profane) for anything but your well-being. Of course when you locate we want to be near you."

His parents remembered that Theo had been enthusiastic about the land around Lake Charm, and they secretly hoped that Dr. Foster could persuade him of the virtues of the Lake Jesup area. However, orange grove land there was expensive, and the Mead properties at Eustis Bluff and Lake of the Woods were only currently disposable at fire sale prices. To buy close to the Fosters would require significant cash. To make sure there was no impediment to such a move if it were to happen, his parents concluded that the ultimate sacrifice had to be made—their beautiful European-finished residence in New York at 674 Madison Avenue was put on the market in early 1886, at an asking price of $45,000.

17.4: Dr. Henry Foster built the Memorial Chapel and Parsonage on the shores of Lake Charm in 1882.

Part Five

The Move to Lake Charm

CHAPTER 18

New Beginnings & Birth of a Daughter, 1886–1887

Another long visit to the Fosters took place in February 1886 and included a tour of Gee Hammock, described in one guidebook as one of the most beautiful in Florida. Theo renewed his enthusiasm for the location, writing, "I am excited over the prospect opened out by these desirable properties at Lake Charm but not so much so I think as to render my judgment bad." He was particularly taken with the returns on citrus shipped that year by steamer to the Northern cities from Foster's Gee Hammock Grove:

> The crop of his groves, 50 acres this year, was 12,000 boxes of which 4,500 were shipped before the frost. He has a contract whereby he got $1.75 a box delivered on the wharf at Lake Jesup. Cost of picking, wrapping and boxing 40¢ tout compris; cost of 24 acres Gee Hammock 10 years with interest at 8% on all expenditures to date $17,000; two crops pay the whole.

But Foster, unlike Theo, was a shrewd businessman. He realized that as far as citrus shipping to Northern markets was concerned, moving perishable fruit by water was ending and the future belonged to the railroads. Jacksonville was over 200 miles by the slow river steamers but only 143 miles away by rail. Accordingly, he paid a bonus of $3,500 to the Sanford and Indian River Railroad to have a line laid to the Gee Hammock Grove, and a further $1,500 to bring the line to Lake Charm and Oviedo, which was completed in June 1886.

Theo's father informed him that they were planning on coming down to Florida for a short visit, and waved aside attempts by Theo to justify the potential purchase of grove land around Lake Charm, "Obviously whatever appertains even remotely to your welfare must be of absorbing interest to us," he wrote. "I do not care a copper whether your calculations of possible profit from the Cater grove or the Jelks grove are reasonable or practicable. It is all-sufficient for me that in your judgment you would be advantaged thereby."

His parents arrived in early March and wasted no time in committing to purchase land for him near the Fosters, as Samuel's diary noted, "March 9: Eventful day. Saw Col. Brewster after breakfast and said we would take the Cater grove, 12 acres plus, in all 36 acres, about 550 trees, some small—name J. J. Cater. Dr. F referred price of front lot to Mr. Lee and settled on $1,750 as compromise, 3 acres with pond near station."

There were no buyers for their Eustis property, jointly owned by Theo and his father, but his father loved the bluff location and had a solution. He offered to buy his son's part of the property at Eustis for his own use for $16,000.

With this money, Theo spent $13,050 on the Cater Grove, reserving the difference for the construction of a home, barn, greenhouse, and improved citrus packing house. Meanwhile, his parents dashed back to New York to take care of the imminent sale of their property on Madison Avenue. They accepted $42,000 cash on April 7, but with the stipulation from the buyer that they would vacate the property by April 26. This gave the Meads only three weeks to move out and find a place to store all their possessions and large pieces of European-style furniture, many of which simply ended up in the auction rooms.

The fourth anniversary of the Mead's marriage was approaching and they remained stubbornly childless. Edith's general health in Florida had been variable and she found the heat very tiring. They decided that Edith would first spend some time with her family at Coalburg before seeing a specialist in Cincinnati, a Dr. Langdon, who was recommended by Dr. Foster. This duly took place over the summer while Theo was in New York, and in answer to a letter from Theo asking about the duration of the treatment, and when they might return to Florida, Edith told him what Dr. Langdon had told her:

> He could not tell though until he saw you what else might be needed. Perhaps he might prescribe a sea-voyage. "For," he said, "what you and Mr. Mead want is something more than a return to health remember." I did remember of course but I exclaimed a little over the idea of a sea-voyage. I am afraid we should have to camp out for the winter in the greenhouse if we went to any such terrible extravagance. Still I know you are ready and willing to do almost anything so that we may have children and if a sea-voyage only will do the business why there it is.

By late August, they were back at Lake Charm, staying in the Foster home, and helping to supervise the building of their new house and the picking of the citrus. Edith's health was fully restored as Theo informed his parents, "She is feeling better and stronger than for years—as though she could undertake almost any fatigue—tho' of course she takes care not to over task her newly acquired strength. It is a great happiness to me to find her so much stronger."

Theo's happiness extended to the packing and shipping of his citrus using the Foster's packing house, and he wrote to his parents that "Altogether I quite feel as if I had sloughed off my Eustis larval skin and was now in chrysalis—chrysalis shell to burst and disclose a 'fly' resident as soon as we get into the new house." Just before Christmas, he shipped 227½ boxes containing 44,164 oranges. The capacity of the packing house, close to the Sanford & Indian River Railroad station at Lake Charm, was about 100 boxes a day, roughly 20,000 oranges, so he calculated that the whole crop could be marketed in 30 days. "It is the nicest way of earning a living that anyone could devise," he wrote.

Word had spread that the Meads were buying citrus land and adjacent grove owners were interested in selling if the price was right. Dr. Jelks owned groves to the west between Theo's land and Sweetwater Creek and offered him about 30 acres of land and 550 trees for $13,000. Theo was all for buying it, telling his parents "It will make a magnificent property. ... if we don't get it somebody else will pretty soon and will hold onto it too," and finishing with a youthful prediction of the financial rewards, "In the worst of years, (barring universal freezes), I don't see how the net return after paying all expenses could be less than 20 percent. ... between us we would have one of the finest, paying groves in the state, good for a minimum of $10,000 net income year in and year out for all time to come."

Theo's father attempted to put the brake on what he saw as an unnecessary gamble, supported by quite a fanciful financial analysis. He likened the proposed expenditure to the threads of a garment that became entangled in powerful and complicated machinery, leading inevitably to one investment inexorably following another. But as usual, Theo's enthusiasm prevailed, and part of his mother's proceeds from the sale of 674 Madison Avenue was used to purchase 32.25 acres from Jelks for $13,200 on December 15, 1886.

His parents arrived late in the year at their Eustis Bluff residence and the event rekindled his father's love of the location, "I shall never tire of the glorious beauty of this place from all points of view. Each seems more admirable than the other," he wrote. Just before Christmas, they helped Theo and Edith move into the new Lake Charm house, although parts of it were still unfinished. They were all kept busy with the hustle and bustle of the citrus harvest, but as recent grove owners, they had to be careful of location when picking fruit. A sharply worded letter from Jelks, who had sold them 32 acres earlier, alerted them that according to grove etiquette although they had bought the groves, that did not include that season's fruit, which still belonged to him.

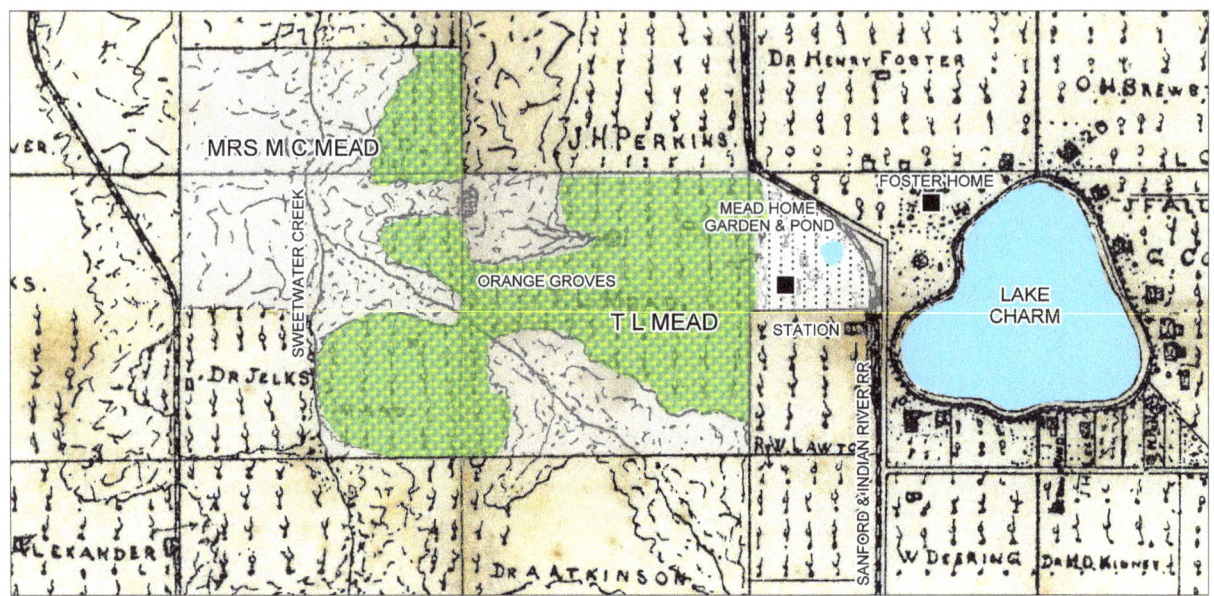

18.1: The Mead land and groves at Lake Charm, shown on a map circa 1890, totaled around 80 acres. The packing house, located close to the Lake Charm station, allowed citrus to be shipped through the Oviedo depot and Sanford.

Picking continued into the New Year interspersed with periods of ditch digging and grubbing up palmettos. Mandarins were scarce that year and in high demand and from the few trees they had they picked the equivalent of eight boxes, anticipating getting $5 to $6 a box. In early March, Theo reported that even though they owned only a part of the citrus crop on their land, as far as early returns were concerned "Eighty-nine boxes sold in Chicago (shipped 97 days ago) netted $149.50—$1.68 a box. ... Sum total so far received (after deducting net losses) $1,537.03 or just about enough to pay the picking, boxing and cultivating for a year."

Theo was elated with the move to Lake Charm, but his mood climbed to new heights with the news that Edith was pregnant. The treatment in Cincinnati under Dr. Langdon had been successful in bringing this about, and after an examination by Dr. Foster they were given a due date for the baby sometime in late September or early October.

The house at Lake Charm, an English-style cottage christened *Waitabit* by Theo, was finished by June. A piano, a present from his parents that Edie bought in a sale while in Cincinnati, was installed in the parlor, which was now ready for the finishing touches. They fitted the parlor picture rail, and one evening as a rest from work began hanging pictures and other wall decorations. European depictions of cherubs and romantic and religious figures of mothers and their children were very much to Theo's taste, and pride of place was a plaster-cast triptych relief of Fiamingo's singing cupids that they hung on the wall behind the piano.

18.2: The newly decorated parlor of Waitabit *with bookcase, pictures on the wall and the piano with the frieze above it depicting singing cupids.*

Since the news of Edith's pregnancy, they had discussed where she should spend her confinement. Edith needed cooler weather and fresh country air away from a city, so they ruled New York out, and Coalburg, which was judged too difficult because of other family pressures. Hunter, Edith's ancestral home, was too isolated for Theo, so they settled on Montclair, New Jersey, a familiar location for the Meads since it was the birthplace in 1823 of Theo's father. It was near Theo's parents in New York and the home of Julia Inness, Edith's long-standing girlfriend. The two had met at Miss Porter's School in Farmington, Connecticut and had kept in close contact over the years.

Theo and Edith left Lake Charm for New Jersey on July 10, leaving Mr. R. D. Barlow in charge of their interests during the period of absence. They rented a house in Montclair until November 9, close to the imposing Inness home of Roswell Manor, which fronted Walnut Crescent. On October 4, 1887, Theo's parents received a telegram announcing the baby's arrival—a little girl. Edith was elated but Theo's initial feelings were of dejection. He desperately wanted a boy, and having not got his own way in this lottery with Nature, had to relieve his negative feelings to his parents in a letter:

> Edith walked and drove as usual yesterday—labor pains began about 5 o'clock this morning and at 11 Dorothy made her début—fat and hearty and weighs 8½ lbs. I think she is a hideous little creature but the nurse says it is a remarkably lovely baby. At present, I don't want to see her or hear her or have anything to do with her, but I suppose I will get resigned to her in time. Of course I pretend that she is quite delightful, but I have to relieve my mind to somebody—*hinc illae lacrimae*. Edie is pleased of course as she wanted a girl. Edie had a very good time indeed the doctor says, she does not seem much exhausted and doubtless will recover her strength rapidly.

Baptism was set for November 5, with the given names "Dorothy Luqueer." From New York, his parents took the Ninth Avenue El to Barclay Street, then the 11:20 train to Montclair, arriving at 12:30. Mr. Carter officiated and the Inness family and another close female friend from Edith's time as a nurse at the Brooklyn Homeopathic Hospital, Miss Camp, was Godmother. A few days later they traveled south, with Edith splitting off the train for Coalburg to spend time with her family before joining Theo later in Florida.

With Edith in Coalburg, Theo was busy with his plants, and lodging locally at Mrs. Todd's family boarding house at Fairview Cottage on Lake Charm—"$8.50 per week and upwards." This gave him time to devote to the sale of hundreds of young Royal Palm plants that he'd grown from seed and were all the rage in South Florida for defining an avenue or growing in tall groves. He advertised them for sale in the December issue of *The Florida Dispatch* under the trade name "Palmetto Nurseries, Lake Charm."

Edith reported news of Dorothy's progress from Coalburg, and in one of the letters told Theo that according to members of the Edwards family, "Dorothy looks like your mother and like you, and has nothing of the Edwards except my ears." It was her mother's view that "she is going to be an engaging little creature", and predicted "that you will be her slave after a little."

Theo had bought a new camera over the summer, and in Cincinnati had stocked up on sensitized plates and paper and a few chemicals for the new process of dry plate photography. He planned to attach a small darkroom to the potting shed at Lake Charm where he would process and print his pictures.

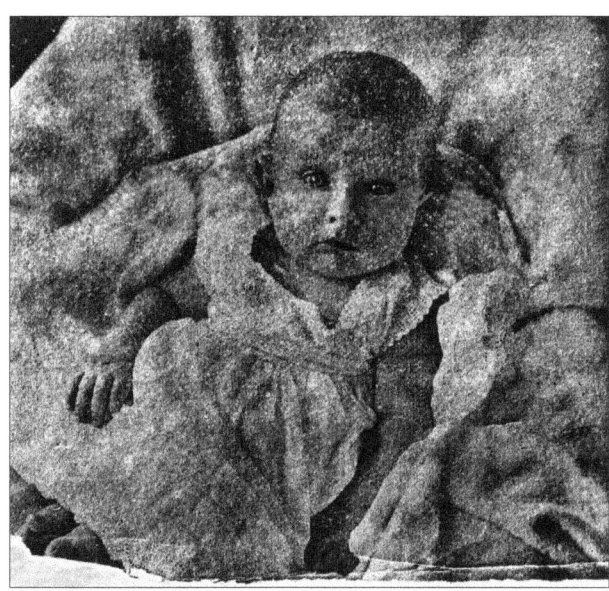

18.3: At Lake Charm just before the end of 1887, Theo made his first portrait of Dorothy.

By mid-December of that year, Edith, Dorothy, and his parents joined him for the Christmas celebrations. The baby became a favorite subject for photography, and, as expected, Theo's paternal attitude began to soften.

18.4: *On Christmas Day, 1887, he took this picture of Dorothy with her grandmother, Mary Mead.*

CHAPTER 19

Improving Lake Charm & Bringing up Dorothy, 1888

The *Oviedo Chronicle* predicted that the 1887–1888 citrus harvest would be a good one for the region. For their part, the Meads shipped 2,097 boxes and by the end of February received returns of $3,350, an average of $1.60 a box. Theo set to analyzing the returns, which varied from $0.73 to $2 a box according to quality, timing and whether sold through a commission house or the Florida Fruit Exchange co-operative. In general, fruit shipped early in the market paid the best prices and the commission houses resulted in the lowest, in part because they took no responsibility for shipping charges or damage, and were quick to reject substandard lots that might have been fine when they left Florida.

Freight charges were a significant part of the grower's profit margin. In late 1887, Henry Plant, owner of the Sanford & Indian River Railroad that connected Oviedo and Sanford, decided to increase the charge on a box of oranges between these two points from five cents to eleven cents. The Lake Charm growers, first by letter and then by a personal appearance of Reverend R. W. Lawton on their behalf, appealed against the increase to the Florida Board of Railroad Commissioners, but to no avail. The citrus growers were furious and decided the Plant system needed competition. Under the leadership of the DeBary Line and Dr. Foster, they formed "The Oviedo, Lake Charm & Lake Jesup Railroad Company" to run track from Oviedo and Lake Charm to Solary's Wharf on Lake Jesup, there connecting with the DeBary Line steamers direct to New York. The major shareholder in the new railroad company with forty shares at $50 each was the DeBary Line, who also committed to rebuilding the Solary Wharf terminal. Among the citrus growers, Dr. Foster secured twenty shares, Theo bought fifteen, W. E. Alexander bought ten, Captain M. E. Brock bought eight, and assorted others purchased between one and five shares each, to produce a working capital of around $7,000.

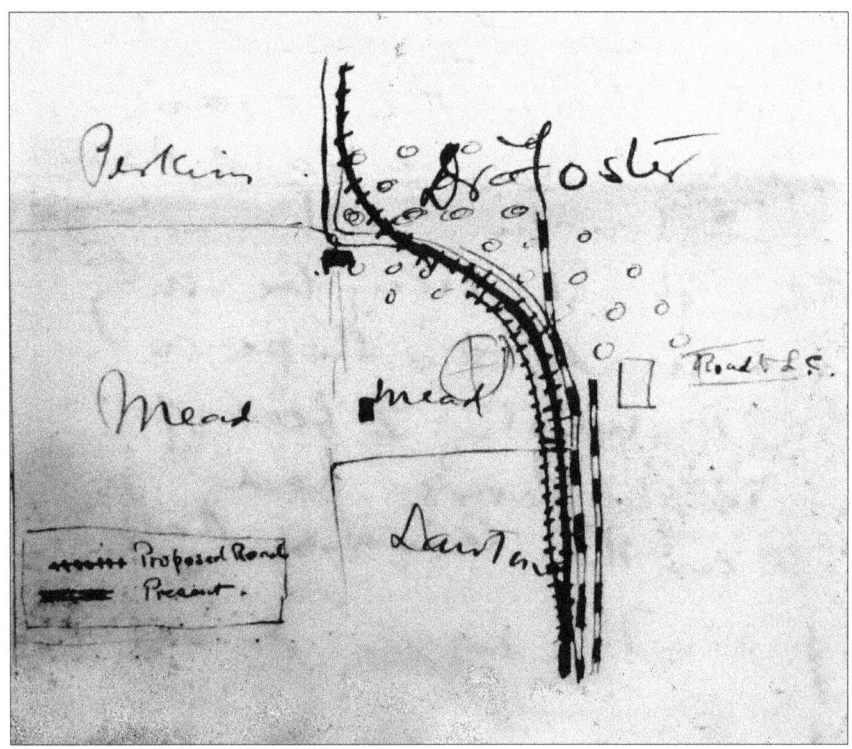

19.1: Theo's February 1888 sketch in a letter to his parents showed the proposed route of the Oviedo, Lake Charm & Lake Jesup Railroad around his property.

Wasting no time, in February 1888, they hired a railroad surveyor to propose a route around the Foster and Mead properties. The preferred line clung close to the existing road to Lake Jesup, but had some disruption to the Mead property,

138   Orchids and Butterflies

as Theo wrote, "It would take a strip 50ft wide off Lawton and our Foster lot and cut a corner of Dr. F's grove off. They say that it would be indispensable for us to have a side track so that to get out we would have to cross 4 tracks, 2 of the new and 2 of the old."

Over the next few months, the company negotiated most of the necessary rights of way for laying the narrow-gauge track, and a hundred tons of light rail were shipped from Philadelphia and deposited at Solary's Wharf. The May 21, 1889 edition of the *Jacksonville Times-Union* carried a story from Oviedo that the last rails for the main branch of the Oviedo, Lake Charm and Lake Jesup Railroad had arrived and that crews were hauling in crossties by handcart to build the tracks. A report in late December 1889 stated that the tracks of the Oviedo to Lake Jesup railroad, now majority-owned by the Clyde Steamship Company, having bought the DeBary interests, were "complete and awaited only the arrival of rolling stock and completion of the rail wharf."

Faced with this determined David, the Plant System Goliath backed down and reduced its freight rates significantly; instead of eleven cents per box to Sanford, they proposed seven cents all the way to Jacksonville, a distance 125 miles further north. The growers had achieved their aim and rapidly lost interest in the new venture, leaving the Clyde Line carrying the major part of the can. Things dragged on until 1892. The task of liquidating the business and tidying up its debt fell to Theo, who spent the intervening years as secretary of the company in this reluctant role. The arrival to Oviedo in the early 1890s of the Orlando, Winter Park and Oviedo Railway, aka the "Dinky Line," brought additional railroad competition and ensured freight rates for citrus would stay competitive in the future.

A further Foster initiative that year was "The Lake Charm Improvement Company." This brought together individual land owners having lake frontage or property close to the lake, who were "desirous of improving and beautifying said Lake, and its surroundings, by planting and cultivating trees, improving avenues, laying out and building walks, and other such improvements, as may conduce to the benefit, health and pleasure of the Community, and to the owners of the lands in the vicinity." Signatures to these objectives came from Theo and Foster and their neighbors, particularly the wealthy Northerners, Calvin Whitney, William Deering and W. S. Farwell, and the hotel and boarding house owners, O. H. Brewster, and J. T. Allison.

The initial objectives were removing mud and grass along the shore and muck from the lakebed, and stabilizing the lake level with a bulkhead and drain. In later years, work would start on building a $1,200 cement walkway around the perimeter of the lake. Assessments for lake frontage were pro-rata according to the amount of shoreline, but generously, Foster agreed that for each project undertaken he would pay half the total costs. Unfortunately, arguments and non-payment of assessments eventually sank this community-spirited initiative.

In the meantime, Dorothy was making up for her early period of little weight gain. Edith had started feeding her mixtures of condensed milk and Mellin's Food made with fresh milk. In February 1888, she weighed 14lb. 8oz. and was gaining about 6 oz. per week. Theo had understandably revised his position of her, reporting to his parents that she had grown a great deal and was "pretty with bright eyes and rosy cheeks." She was however still a little colicky, was going through her teething period and had a nervous disposition, jumping at loud sounds like a slammed door.

As summer approached, it was decided that Edith would take Dorothy, variously nicknamed "Dar" or "Polliwog" (shortened to "Polly" or "Wog"), and seek the cooler climes of St. Augustine, leaving Theo to get on with planting peach trees in the orchard and spreading muck as fertilizer on the pruned citrus. By late June, they were back at Lake Charm and Theo reported to his parents, "The baby is getting on very well, she doesn't have quite as good color as at St. Aug. but seems quite well and good natured and active—only howls when put in cradle to go to sleep." Her teeth were coming through—the seventh was just showing—and sucking on a hard blunt object too large to swallow gave some relief.

19.2: Dorothy in her carriage in the garden at Waitabit, *July 18, 1888.*

In late October, Edith decided she must take Dorothy to Coalburg, a brave decision at the time since yellow fever had broken out that summer in the city of Jacksonville, creating a state of panic that spread to Central Florida. Many of the inhabitants there had fled the city, while those left behind resorted to burning tar pyres in the streets to "purify the air." The authorities introduced strict quarantines, the disinfecting of goods by burning or fumigation, and the firing of heavy cannons, believing that yellow fever microbes were killed by concussion from the sound.

19.3: Theo's delight in having a child of his own comes through in this photograph of him holding Dorothy, taken at Lake Charm in 1888.

Although confined to Jacksonville, there was mounting concern among Floridians who had to travel about running the gauntlet of the quarantine set up around the city, since almost all north-south transport passed through the area. For through passengers, railroads had adopted a strategy of running special non-stopping quarantined trains around Jacksonville, the trains starting from Orange Park south of Jacksonville and running to Waycross in Georgia. All mail to and from Florida was sulfur-fumigated at Waycross, the belief being that fumigating all mail would be faster than separating mail individually from infected areas.

Edith and Dorothy left Lake Charm on October 26, and she wrote to Theo of her safe arrival at Richmond on October 28:

Here we are all safe and sound having arrived at 7:15 a.m. instead of in the evening as they told us at Sanford. We found the sleeper at Waycross and again I had to lug all the things except the bag of diapers. From Orange Park 14 miles south of Jax to Waycross we were in an ancient and disgustingly filthy car and all the windows were kept shut and doors locked. The floor was covered with tobacco juice and at every seat was a nasty looking spittoon that had not been cleaned all summer I should say. The air was redolent of tobacco juice and sulphur and made me feel faint. ... It did seem good to get out of that vile evil smelling car into the handsome clean Pullman.

At Coalburg, Catherine, Theo's niece and now almost five years old, wanted desperately to be friends with Dorothy, who was having none of it. Dorothy's nervous disposition seemed to be aggravated by strange surroundings and people. Edith reported to Theo, "Dorothy has been in such a nervous excitable state since we came that I have had to hold her in my arms all the time. If I put her on the floor a moment she screams herself black in the face and as to going to anyone, she repudiates the entire family even the little cousins." The testing behavior continued with Edith complaining, "I am fairly at my wits end with her" and, "They all think her utterly spoiled and unmanageable." There was a definite battle of wits going on.

All the family were concerned about the possible spread of yellow fever to the Central Florida area, including Catherine, who was overheard at the end of her prayers saying, "God bless Aunt Edith and Dorothy and don't let Uncle Ted get yellow fever." She commented the following day, "I am sure Uncle Ted would send me some oranges if he knew I prayed for him every night." To many the power of prayer appeared to have worked when on November 25, with the first frost of the season killing the mosquitos, the Jacksonville epidemic disappeared as quickly as it had arrived.

CHAPTER 20

Bumper Citrus Harvests, 1889–1891

In early 1890, duty called on Edith again to dash back to Coalburg, leaving Dorothy with Theo, following news that her mother was ill in bed with a severe attack of the whooping cough. Left alone with Dorothy, and Louise the servant girl, Theo got less work done around the groves but spent valuable quality time with his daughter. For relaxation, he decided to take some pictures of her in classical and unusual poses. In one he had the Greek bronze vase, given as a wedding present from his Alpha Delta Phi brothers, as a prop with Dorothy dressed in a Greek-style robe; in another she was photographed emerging from a large woven Alibaba basket. Theo sent a print of this second picture to Edith at Coalburg, who was delighted with the result, writing, "Have yours of 6th with the sweet picture of my own Jill-in-the-Box. How much obliged I am to your trouble you were at in taking it for me. It is ever so sweet and all admire it here."

Edith, torn between a daughter's duty to her mother and her maternal bond to Dorothy and love for Theo, longed to return to *Waitabit*, and wrote, "It has been a sore trial this parting and no one appreciates it or seems to think I have any just causes for returning home." Edith's mother was on the road to recovery by late February, and Edith was finally able to persuade the family to let her return to Florida.

20.1: *Dorothy photographed by Theo in early 1890, as a Greek muse posing with the Alpha Delta Phi bronze vase and emerging from a large woven Alibaba basket.*

Back at Lake Charm, the citrus returns for the 1889–1890 harvest were all in, and with the usual liberal application of fertilizer, the groves had been productive and were now in excellent condition. In March, new growth covered the trees with a mantle of green and the orange blossoms filled the air with their scent. With good market prices over the winter, their total net income was $5,700 on 3,700 boxes shipped, providing the Meads with more than $100 a week, enough to make Theo independent of family money for support. The years around this time would prove to be bumper ones for citrus—in the 1890–1891 season they shipped 4,600 boxes and pocketed $7,200.

Over the years, individual trees varied in their citrus yields, with the larger and older ones providing more than a thousand oranges per tree. In 1886, he had recorded, "We have picked from the 44 trees 54,126 oranges or an average of 1,230 from each tree, or 6½ boxes." One tree in particular, that was named Catherine's tree after his Coalburg niece, was singled out for its prodigious yield, "One of the men remarked concerning one tree (that he had then nearly picked) that he never saw such a tree for oranges, it beat any of those on Dr. Foster's grove, it was covered all over, top, bottom and sides. He picked 18 or 20 scant boxfuls—equal to 12 or 13 packed boxes, probably 2,000 oranges from it."

Picking crews of as many as eight men were organized under Gus Nelson's direction, working 12 hour days and capable of picking upwards of 80 boxes per day, containing on average 200 oranges per box. The labor bill for 80 boxes was 105 hours @ 12½ cents plus $1.50 for Gus' supervision, or 18 cents per box. The pickers left any oranges that were not fully ripe and marked with chalk on the tree trunk the approximate yield of each tree.

Theo's modern packing house, close to Lawton's grove and the station, was up to the task of packing and shipping these large quantities. A man at a separator initially divided the stream of oranges according to their exterior appearance, bright orange or russet-colored, and these two streams were then sized automatically and sent to bins. Within each size, the bright orange-colored fruit were further graded and packed into gilt-edge, fancy and extra, and the russet fruit into light russet and russet to which Theo added "golden russet," which were oranges very black with a bright orange cheek. These brought the same price as "extras."

The Oviedo Chronicle of February 1889 was a friend of the russet, reporting, "The people of the North are finding out that inside the dark skin of the 'African' russet orange is contained as fine a flavored fruit as can be found anywhere, and the consequence is that price of 'Africans' has been steadily growing better."

20.2: Top left: Gus Nelson and his picking crew; Top right: Three pickers at work on Catherine's tree; Bottom left: The packing house; Bottom right: An aerial view of the packing house looking according to Theo, "like a giant spider."

Fertilizer remained a significant cost item and was essential to achieving good citrus yields. Some muck was available on Theo's land and that was supplemented with Mapes-brand "Fruit and Vine" fertilizer, but Theo was keen to remove costs by compounding his own out of a 50/50 mixture of hull ash and bone meal, reckoning that he could duplicate the commercial composition and save at least $10/ton. Each year he was fertilizing at a rate of 25 lb. per tree requiring more than 15 tons of fertilizer for his groves of approximately 1,200 trees. He ordered 7.5 tons each of ash ($29/ton) and bone ($36/ton), calculating a cost at $32.50/ton against $43 for the Mapes brand.

With good money coming in from citrus, and with the improved grading and gravelling of rural and primary roads in the Oviedo area, Theo ordered a phaeton from Wilbur Murray of Cincinnati for $86 to provide covered horse-drawn conveyance to the local stores, church on Sunday and for calling on friends.

At home, the battle of wills with Dorothy was in full swing and she had now entered the "terrible twos" phase. By the start of May 1891, and with the temperature climbing, things were getting so bad that it was decided that Edith and Dorothy should both go north to cooler climes. Edith had a smattering of French and German and wanted to improve her language skills. Together they built an itinerary for her around a summer school of languages at Burlington, Vermont, visiting her friend Julia Innes at Montclair and Theo's parents in New York, and ending at Coalburg where Theo would join them later on. Travel was booked on the Clyde steamer *Iroquois* leaving Jacksonville bound for New York on July 2. Theo was characteristically blunt about the arrangements, stating, "I shall be glad to pack both of them off."

Edith's trip north got off to a disastrous start. She arrived at Julia Inness' home of Roswell Manor via New York, only to have Dorothy go down with suspected chickenpox. Edith sent the news to Theo, "Dorothy is not so well this afternoon, has eaten no dinner and lies on the bed in a drowsy condition, though she only slept once for a short time. If by tea time she is no better we are to have the doctor." A likely case of whooping cough was diagnosed and bed and rest for several weeks recommended. Dorothy was contagious and coughing badly, so traveling anywhere else was out of the question. Attending the summer language school now appeared to be a lost cause, and Edith was tired and exhausted. Dorothy's emotional state had been a worry to Edith all summer, and she reasoned that the sooner she got to Coalburg the better. Once there she hoped that Catherine, a very gentle child, could be Dorothy's associate and help calm her down.

With Edith and Dorothy away, Theo was "monarch of all he surveyed," and with the summer of 1891 ahead of him he intended to do exactly as he pleased, but he certainly would not be idle. His groves and fruit trees and bushes needed constant attention, as did his greenhouse plants, and there was always weeding to do. Also, orchid growing was beginning to be of serious interest and in July he reported to his parents:

> There are three handsome orchids out now—two laelias and a cattleya. Have hybridized a pod of *Cattleya leopoldi* with pollen from *Laelia anceps*—the union is shown by the petals of the flower quickly withering away, while the rest of the flowers remain fresh. I have also another hybrid pod—*Cattleya schilleriana* x *Laelia purpurata*—when the seeds ripen I will get a great slab of royal fern root, I can get them here a foot thick and high, and dust the seeds over it and leave them to their fate—treating the fern root just as I would if an ordinary cattleya were growing on it.

By late October 1891, Theo figured it was time he joined Edith at Coalburg. When Theo arrived, he found Dorothy to be more robust than he'd seen her before and happy with Catherine, singing and playing all day long. "I don't know what she will do when she gets back to *Waitabit* and has no constant playmate," wrote Theo to his parents.

Edith's sister Anne, Catherine's mother, had lost her second daughter Emily in April 1890, aged two, and was combing the area looking for an eligible child to adopt. Theo's characterization of this search was, "The demand for babies seems to be brisk and prices firm with an upward tendency—stocks on hand firmly held in most cases with no concessions to the purchaser."

20.3: Dorothy (aged four) and her cousin Catherine (almost seven) were constant companions and playmates at Coalburg in October 1891.

Edith and Anne visited an orphanage at Charleston and were both smitten by a skinny two-year-old boy with fair hair and a light complexion, who had been abandoned by his parents. Anne pointed out that he could be an excellent companion for Dorothy, just like Catherine, and also fulfill Theo's desire for a boy. Edith and Theo were eventually persuaded and agreed to take him on approval for a year. Theo reported the decision to his parents, "He is picking up very fast, but Edith says would certainly have been dead by this time if he had remained in Charleston. Edith is getting very fond of him already. He seems a rather impassive and phlegmatic temperament so far, but will probably be more frisky when he gets a little more flesh on his little bones."

Theo was concerned about the effect on Edith's health of looking after two young children. He told her he thought it might be too much physically for her, and was perfectly willing to return the boy tomorrow if necessary. Anne had quite fallen in love with him and Edith agreed to try it for a year. They decided to name him Harold, but Theo had one final say before they left Coalburg in late November to return to Florida, writing, "If at the end of the year he were not thoroughly healthy and satisfactory I should have no hesitation in returning him, even though we should be fond of him personally."

In early December at Lake Charm, things were not going smoothly. Theo had the grippe with periods of neuralgia and Harold had a prolonged attack of diarrhea. The washing of bedding daily contributed to Edith's already heavy workload. There was no harmony either in the relationship between Dorothy and Harold, prompting Theo to unburden himself to his parents:

> Nobody was ever sicker of a bargain than we of ours—the "Home" at Charleston will be in despair but we can't help that. We shall put the main ground of his rejection on account of the ill will he constantly shows towards Dorothy—he snaps and snarls at her constantly tho' she doesn't tease him. Nothing seems to remedy his bowel complaint and he refuses to give any notice of his needs at any time tho' he will demand anything he wants fast enough.

All the rooms had to be thoroughly cleaned and tidied for Christmas, including Theo's den. Amid much grumbling, Edith asked him if he didn't think it was worthwhile to get rid of all the dust and cockroaches. Theo assured her that "nothing whatever was 'worthwhile' under existing circumstances." Overall they were both feeling very down and dispirited but as the New Year of 1892 arrived, there would be much worse to come.

CHAPTER 21

Scarlet Fever Strikes, 1892

On February 8, 1892, Dorothy came down with a sore throat and high temperature and had to be put to bed. Dr. Foster diagnosed scarlet fever with typically a six-week illness period. Persuading her to take fluids with her throat hurting continuously was a frustrating business, as Theo reported to his parents at the Bluff:

> Her throat seems to hurt her at times and she won't take either food or medicine as a general thing—once she took a sip or two of milk—about the only food in 24 hours and said "that milk is so good, Mamma!" But she wouldn't take any more. One time she was crying out for water but set her teeth and absolutely refused to take it when offered though all the time begging for it.

By February 13, Dorothy appeared to be making some progress in recovery although she was still refusing nourishment and sleeping most of the time. By February 15, her fever had abated somewhat and she was taking a little more milk. "We are as anxious as it is possible to be and can only await results," wrote Theo.

Theo had a good working knowledge of illnesses and diseases, having read widely on the subject, and he joined with Dr. Foster to discuss Dorothy's prognosis. They agreed that the chief worry was that the toxins produced by the infection could affect other vital organs, such as the kidneys, heart or eyes. But for the next few days there continued to be a slow improvement; she was able to breathe through her nose rather than her mouth which helped her sore throat a little, passed water and was well enough to be a little cross one afternoon which was interpreted as a good sign.

On Wednesday, February 17, Theo reported further encouraging signs, although she was still spending most of the time in lethargic sleep:

> Dr. F came today at noon and said that Dorothy was doing fairly well, considering all things. She said "yes" in a weak little voice when he asked her if she felt pretty well, showed her tongue on request and answered "goodbye" to his goodbye—more signs of recognition than we have seen for a long time before. Unless roused for the moment though, she continues in her lethargy—all the other symptoms however show considerable improvement—throat and nose, etc. and she opened her eyes a few times.

Looking after Dorothy was taking it out of both of them, particularly Edith, although her Aunt Mary was on hand to relieve them while they both tried to catch a few hours' sleep during the day. The medicines were so numerous and varied that they had to make out a regular timetable for the nighttime. For the first half of the night, every drop swallowed produced a severe cough, and it often took five minutes and more to get each teaspoonful down. Towards morning, the cough lessened and it was a little easier to administer the drops although she still would not open her mouth, and the medicine and milk had to be allowed to trickle in between her teeth. Theo continued to sit with her for most of the night and on February 23rd, it barely registered that that was his 40th birthday. With all the time needed to care for Dorothy, Harold's fate was to board temporarily with the Christiansen family in Oviedo; honest people who ran a nice clean place with lovely flowers and who agreed to take good care of Harold for $3 a week.

Dorothy was doing "as well as might be expected" according to Dr. Foster—there were one or two encouraging signs, but fever and coughing at night and the battle to give her medicine or nourishment orally continued. She could move her head a little from side to side but expressed terror whenever the atomizer came out, and jerked her head away whenever a dropper full of medicine, milk or water came near, pain getting the better of hunger or thirst.

On February 24, Theo's progress report to his parents was uncharacteristically downbeat, almost as if he was on the point of abandoning all hope. He wrote, "Dorothy's condition remains about the same—she is very nervous and acts like a frightened wild animal when fed or sprayed—Dr. F was over this morning and said to give up spraying and squirt a little of the antiseptic into her mouth when fed or medicined as it would not interfere with digestion if swallowed."

Dorothy was getting very much weaker. The glands on her neck were large but not large enough for lancing, and Edith had to cut her hair short because she could not brush it without hurting and worrying her. The progress of the illness was slow yet remorseless and the battle with the toxins attacking her vital functions was essentially lost. Darkness closed over her in the early evening of Thursday, February 25, and she slipped away. Theo sent a telegram to his parents at Eustis Bluff that was delivered by a horseman at 9 p.m. It read simply "Dorothy died 6:15 p.m."

The next day his parents rose early at three o'clock from their home at Eustis, took coffee and made their way through a cold fog to Tavares, where they boarded the Florida Central & Peninsular Railway to Orlando. After a two hour wait, they made the connection with the Orlando, Winter Park & Oviedo service, arriving at Lake Charm at one o'clock, and gave what comfort they could to the distraught parents, bewildered and shocked that after only seventeen days and nights she was gone.

Sanford's oldest undertaker, T. J. Miller, was contacted, as was the Reverend Lyman Phelps, who was asked to perform a burial service the next day at the Mead property. On Saturday, February 27, with Phelps officiating and led by

Miller, Messrs. Storrs, Clatworthy, Armstrong and Couch as pallbearers bore the casket by its handles around the newly emerging starry-blossomed fragrance of the orange grove and back to the house, where the assembly rested and sang "Asleep in Jesus." There were yellow roses from the Phelps' son and a basket of flowers from the Todds. Dorothy was buried in her own little patch of garden close to the house, where Theo later planted a tree azalea near her grave. On the following day, Sunday, a service was held in the Lake Charm Chapel led by Dr. Foster.

21.1: Believed to be the last formal picture of Dorothy, probably taken shortly after her fourth birthday in late 1891.

Condolence letters flooded in following Dorothy's death. Theo's mother gave her reaction to the tragedy, "Our precious one has been nearly a week in glory when you receive this. Let us join in her happy song too, Glory to the Lamb who was slain for us!" His father wrote on the bottom of the same letter, "Should like to see Dorothy's garden—keep the darling's photo in sight all the time, the full face one with sprays in lap. No mischance can take away the four years we had of the delightful darling."

21.2: Theo's father counseled him after Dorothy's death to keep this photograph close at all times.

The month of March was marked by long periods of stasis for both Edith and Theo. They walked the circuit of the groves every morning after the usual daily renewal of the roses in Dorothy's garden, and swapped anecdotes about Dorothy and some of the funny things she would do. Edith remembered her love of the smell of cocoa butter, on one occasion taking a small cake of it to bed with her, which ended up getting everywhere; Theo, the time when she got a notion of taking a small traveling clock with her to bed and hugged it close to her saying that it was "such a comfort." Theo did some odd jobs about the greenhouse but was mostly alone with his thoughts. At Sunday church, as he listened to the sermon from Bishop Foster, the full extent of his loss dawned on him as he recounted to his parents, "I was as still as a mouse all thro' the long sermon. It seems to me for the most part as if the darling had gone on a journey and might come back any time—she was away from me more than a quarter of her whole life anyway but Edith never was separated from her except for the six weeks when she went to nurse Mrs. Edwards through the whooping cough."

Harold was collected from the Christiansens and he turned up rosy and well, and pleased to see them both. Edith felt the sadness of seeing just him without Dorothy, but there was some comfort in his hugs.

Although Edith was approaching forty years old, after a full medical examination by Dr. Foster, he informed her that there was no reason why she shouldn't conceive again. He gave directions to both as to diet to build up general robustness. Theo was desperate for a family and wrote to his parents that it would be "the only thing that would make life really worth living." His father expressed surprise at this statement, pointing out that many people would think this a marked idiosyncrasy and unusual, believing that for Theo it was caused by "want of free intercourse in early life with other children of same age."

There were few bright spots for the rest of 1892, but one happened in June when Theo received a letter from Harry Robie, informing him of the birth of their second child, a boy, whom they proposed to name "Raymond Mead Robie" in his honor.

Over the summer, Theo's parents were going North to the Chautauqua meeting at Lake Chautauqua in New York State, and offered the Bluff house as a change of scene. Theo and Edith declined the invitation, having already decided to spend the summer around Linville, North Carolina, where they could board for $15 a month. They were looking forward to the bracing air of the mountains, and a rest period spent walking, relaxing, and reading. After a couple of weeks there, they had come to a final decision over Harold's future that Theo communicated to his parents:

> I am not satisfied with Edith's rate of improvement here and know that the care of Harold is too great a strain on her. Moreover the boy does not seem to have advanced mentally a particle since we have had him, and after somewhat prolonged consideration I think we had better take him whence he came.

In mid-September, they handed Harold back in Charleston, as Theo described:

> We brought down Harold today and left him at the home—a clean pleasant place, well-kept where he will have all the comforts he needs. Mrs. Ruffner who is president of the association is a very kindhearted lady and says she will try and find a home for him with some plain-food Christian people of the miner class—they get $3 or $4 a day wages and can bring him up very well. She thinks it wouldn't pay to waste a good education on him. She says she would not like to see him go back to his father who has just married some low-down cracker woman.

News from Florida in letters arriving at Coalburg was that his plants were flourishing, but lubber grasshoppers had got into the greenhouse and done considerable damage to the orchids. Theo felt as if he would burst if he couldn't get back immediately, so by October they were back at Lake Charm and Edith flung herself into housework and the preparations for Christmas. This was always an active time for the Meads who not only sent out dozens of Christmas cards, but also embraced the giving of small gifts to all the surrounding children, particularly those from poorer families. Theo's letter of December 20 summed it up, "Edith is overwhelmed with her Xmas meal and works on tho' ready to drop. She has provided for forty of her friends and relations besides nearly as many negro Sunday School children and is in despair because she can't make things for the remainder of the human race before the day arrives."

The Meads spent Christmas Day at Eustis Bluff with Theo's parents in their newly rebuilt house, which Theo described as "quite convenient and pretty." On the way back from Eustis, they called in at the Robies at Mount Dora. Harry was putting in a windmill tank 50 feet high holding 17,000 gallons of water for citrus irrigation, but he had a unique and very personal present for them. Having become tired of the first Christian name "Raymond" for their second child, they told him they were changing it to "Theodore" in his honor, following the loss of Dorothy, and would he consent to be Godfather to "Theodore Mead Robie"?

The New Year of 1893 entered with a whimper and brought to an end the Mead's "annus horribilis." They received a letter from his mother ending by wishing Edith the happiest year of her life, to which Theo replied, "Edith says the happiest year of her life will be the year when she sees Dorothy again."

CHAPTER 22

Irrigation & The Great Freeze, 1893–1895

Since moving to Lake Charm, Theo's citrus trees had all been hand and hose-watered in the dry season from drilled flowing wells. In 1891, he cooked up an ambitious citrus irrigation project involving the pumping of artesian water by steam power to irrigation heads close to the trees. Work started in the summer, drilling the flowing wells near the workshop engine house where he planned to locate the pump. He bought a boiler and second-hand Worthington steam pump in Sanford for $300, and employed a mason to build the foundations for the pump and boiler in the wooden workshop. By September, he had spent nearly $2,000 on materials, equipment and labor and over the next six months spent a further $660 on additional items.

To complete the project, main pipes and branches needed to be laid and he estimated it would take 2,000 feet of pipe costing 60¢ per foot to reach as far as the Sweetwater groves. The system, completed over the summer of 1892, was sufficiently novel that it made the local newspaper:

Mr. T. L. Mead, the famous florist, of Oviedo, has just completed a splendid irrigation plant at his place on Lake Charm. He has a six-inch artesian well that brings the water within thirteen feet of the surface and is pumped up by an engine in any quantity desired. It is conveyed to his grove and garden by pipes. He has the contract of irrigating the groves of some of his neighbors.

22.1: Mr. Edgar (left) was contracted to drill a number of flowing wells to supply water to Theo's steam-driven pump.

Theo demonstrated his citrus irrigating system to Dr. Foster and impressed him with an exhibition of waterworks that included using a single hose to blast away sand from the collar of a tree. Foster asked him to get him two long linen hoses and contracted him to irrigate the trees in his local grove at a pay of $1 an hour. Robert Lawton, with his grove just to the south of Theo's, was also impressed and wanted a similar service for which the pay would be $1.20 an hour. Theo reached an agreement to irrigate Foster's groves on Friday and Saturday, and Lawton's on Monday and Tuesday. Theo reckoned that he could count on taking around $400 annually in reasonably dry years, so long as "the clerk of the creator doesn't take the job off my hands as happened twice before."

22.2: Theo in his boiler house tending to a gauge on the Worthington duplex steam pump. Note the rocking chair, newspaper and folding bed for nocturnal pumping to protect citrus when temperatures plunged.

With the stress on the Mead family finances still present, keeping busy like this and earning some cash to keep things running was valuable, and took Theo's mind off the distress that they both felt in losing Dorothy. Edith had no such outlet, but Theo was always supportive. Despite her relative youth, Edith's frequent headaches didn't need much to trigger them, as Theo recounted, "We went to chapel this p.m. and heard the new inmate of the parsonage, Mr. Shove, quite inferior to the last batch of people Dr. F had here but about what we expect of a rural Methodist preacher—doubtless a very estimable person in his way. He was long-winded and naturally gave E a headache."

They were still desperate for a family. Theo believed that their chance of further offspring was small and with time being short, they ought to consider living in the North for a year, even under forced economy. He expressed his thinking behind this conviction as, "I think the climate of Fla. is largely responsible for the sterility so common among Northern women who live here." He added, "The prospect of leaving my orchids and things is anything but agreeable, but there is no sacrifice that I would not make to have a child of my own."

Instead of living in the North, however, Edith was packed off first to her family at Coalburg, and then for an extended course of treatment with Dr. Foster at Clifton Springs in an attempt to improve her health and the chances of conceiving again. Edith was in Coalburg for the month of June before a four-month stay at Clifton Springs, the purpose of which mystified her mother as Edith reported to Theo, "Mamma takes it hard, and can see no earthly reason why I need to go there. She would laugh at the true reason so I don't tell her."

At Clifton Springs, Dr. Foster conducted a full examination and stipulated an extended course of stimulating baths, massage and electrical treatments. She was required to lie down and rest after each procedure and be out of doors all the rest of the time when not eating, so was fully occupied. Edith was homesick and probably bored with the repetitious regimen, and wondered in a letter to Theo, "Is it worthwhile, when life at best is so short and we have already lived such a large part of it, for two people who love each other to be parted so? Today I don't feel as if anything—except a baby—was worth such a sacrifice."

After a couple of months there, her Aunt Mary commented that she was "looking like a new woman since you came here," and Edith told her she felt like one too. Dr. Foster was sure she would leave there strong and well, provided she could stay long enough, until October at least.

With Edith away over the summer, Theo was busy irrigating his and his neighbor's citrus trees and when the rains came, spending more of his time growing and hybridizing orchids. The garden was increasingly devoted to the growing of edible fruit and vegetables, providing food for the table that took little time to prepare. He kept up his letter writing to his parents and to Edith during what little leisure time he could find, even if this were a Sunday. This frequently brought a sharp rebuke from his mother, as in this example, "It is a very interesting letter but it was written in the Lord's time, and I fear you will fail in receiving the blessing pronounced upon those who remember His commandments to do them."

Theo joined Edith at Clifton Springs in October, and managed a side-trip to Ithaca to renew his friendships with his Alpha Delta Phi brothers, telling his parents, "I guess there aren't any adjectives in the language adequate to say how good a time I had with the boys yesterday—it seems as though it might be enough to spread over the whole year or two that may elapse before I get back to them again."

22.3: Theo had an early interest in cacti and succulents dating from his Cornell days at Ithaca.

164   Orchids and Butterflies

By November, they had waved goodbye to Clifton Springs and were back at Lake Charm. In their absence, the caretaker Ward had taken good care of things barring some inevitable orchid damage from cockroaches in the greenhouse. Edith, who had stood the journey well without headaches, began the work of cleaning the house, but confessed to Theo that her hardest trial was the silence and loneliness of the rooms without the sound of Dorothy's voice.

Theo had always found gardening to be a great comfort in times of trouble, and as a way of helping to deal with his grief had retreated more and more into hobby aspects of horticulture. He had purchased his first succulent in Ithaca, shortly after arriving in 1875, and had quickly become fascinated with their water-conserving properties and began collecting them. In his rooms at Ithaca, there were pots of *Schlumbergera*, the Lobster or Crab Cactus and an extensive collection of smaller potted varieties. At Eustis Bluff, he had expanded his collection to include larger versions of yucca and agave and had brought several of them to Lake Charm, including a specimen of *Cereus*, which he planted growing up a palmetto on the south side of the house. In May 1894, this rewarded him with a spectacular night-blooming display of sixteen flowers, which he rose at five in the morning to photograph.

Earlier in the year, in May 1894, panic had gripped the stock market caused by the overexpansion and shaky financial status of railroads, which led to several bank failures. There was a run on gold and a severe economic depression spread across the country, accompanied by a decline in prices for all agricultural commodities, including citrus. The orange harvest at Lake Charm was abundant, but the bottom had fallen out of crop prices and returns were barely enough to cover the cost of growing, picking and shipping. At this point for Central Florida citrus growers, the Gay Nineties were anything but gay. Compounding the Mead family's troubles was their growing cash shortage, and Theo's father reluctantly had to turn to his New York property portfolio. Properties were owned and rented out in Manhattan on Greenwich Avenue, at Perry Street and Waverley Place, and at the corner of Greenwich and Charles Street, which was placed on the market. "If we don't sell Greenwich and Charles we stand a chance of being busted next year," commented his father. The eventual sale of this property averted a financial crisis, and the orange crop from Eustis Bluff that year raised a further $1,500.

As the weeks and months of 1894 flew by, Edith and Theo slipped into their regular routines, with days always ending with Theo reading to Edith by the light of the kerosene lantern before lights out at nine o'clock. The piano, a housewarming present from his parents when they built *Waitabit*, was getting regular use too, with the bundles of sheet music his mother had sent with it. Edith set herself the target of playing a new Beethoven sonata every week, just to practice her sight-reading and found this relaxation a great comfort.

By December 1894, Edith was preparing to leave Florida again, this time heading back to Coalburg to look after her mother who had fallen ill with a severe case of whooping cough. On arrival, she found her mother already feeling better, so she had no need to stay for long once Christmas was over. She wrote to Theo's mother, "The weather is turning colder now and snow is beginning to fall, much to my joy. I do so love snow and real winter weather."

At Lake Charm, winter weather was the last thing citrus growers wanted but it was what they got. With Edith still at Coalburg, on December 29, 1894, a severe freeze blanketed the State, bringing sub-32 degree temperatures to nearly every part of Florida. The minimum temperature was 14 degrees at Jacksonville and 19 at Ocala and Tampa. The following night was similar and apart from Key West, the rest of the State had minimum temperatures below 26 degrees. It struggled to get above 32 degrees during the day, and all ungathered fruit were frozen on the trees or else had fallen with the leaves to the ground.

In Central Florida, there was widespread disaster with young trees killed outright and the tops and smaller limbs of old trees blackened and severely damaged. The winter crops were completely ruined. A contemporary account of the calamity reported, "Everyone feels about as blue as can be. One of our neighbors at the sight of his ruined grove took a chill and went to bed sick."

22.4: *The Great Freeze of December 1894 and February 1895 killed almost all the trees, and brought commercial citrus growing to a halt for the next ten years in all but the southern parts of Florida.*

Growers had already shipped some of the early crops, but an estimated 2.5 million boxes of oranges, valued at more than $2 million, were frozen solid and had to be destroyed. There were complaints from growers that the Weather Bureau had underestimated the degree and extent of frost, since many took no action believing that their trees would survive what had been forecast. Nevertheless, there was a glimmer of hope that not all was lost. While fruit and younger trees were ruined, many of the older, better-rooted trees reportedly survived and only seemed to thrive after the severe cold snap.

Edith returned from Coalburg in the first week of January, which turned out to be a warm, wet month encouraging the old trees to produce sap and begin growing again. On February 7, things got worse as a further arctic blast swept the State. Accompanied by a strong wind, it brought a vicious return to freezing temperatures. It was 16 degrees in Tavares and 18 degrees in Sanford on February 8, and for forty-two hours the mercury stayed below freezing. The rising sap expanded as it froze, shattering the old trees from the inside out. People said that during the night they heard the sound of the bark opening on the trees as if someone was outside cracking walnuts. The next day icicles of all sizes decorated the trees and bark hung in sheets and fragments around the trunks. Some of the larger trees split wide open and their roots froze. Theo reported to his parents:

> It has been blowing great gusts today from the west and Dr. F. says barometer is falling unusually low—it is getting colder too and reports for the North are of intense cold. Another bad freeze would annihilate the groves at this time. The blighted trees that I moved into the cow lot seem to be dead from the freeze—they had pushed out a few tender shoots after being moved. They were big trees weighing a ton a piece and so the frost caught them in a susceptible condition. The bark is cracked off their trunks in places.

Those that had the money bought grove land further south and started all over again, but many simply gave up on Florida and headed back North, their hearts filled with melancholy, as one observer reported:

> I stood upon the bank of the lake and watched the wagons filled with sorrowful-looking men and women on their way back North. They had built their houses and made their groves and then saw them swept away in one night by the cold winds of the northwest. They had risked all and lost and now they were abandoning what was left of their once beautiful homes.

With all hope lost, stories abounded of people leaving their properties in such haste that unwashed dishes were left, together with linens and all kind of household equipment. It was estimated that one-third of the people of Florida had already left the State, many more were planning to go, and even more would go if only they could.

The impact of the Great Freeze on citrus production and on the livelihood of growers was catastrophic. Not only was the major part of the crop lost for one year but the trees as well, representing future production for the next decade. From a shipped harvest of five million boxes of fruit in 1893, the State only managed 147,000 boxes in 1895, a reduction of 97%. Overall aggregated losses were estimated at $15 million. As for the Meads, they saw their income drop from an average of $100 a week to just a handful of dollars. The growers found little comfort either in the fact that at least they still owned the land. Before the freeze, the asking price for desirable orange grove land was around $1,000 an acre; after the freeze, desperate sellers were lucky to get $10 an acre. In June 1895, the 120 acre Markham grove of mature trees about halfway between Sanford and Tavares, together with its $6,000 irrigating plant, was on the market for $5,000. This sale price was the final insulting conclusion to the tragic story of Mr. Markham, who earlier on the morning of the second freeze had gone out into his groves and shot himself.

Part Six

Everyday Life in Central Florida

CHAPTER 23

Humanitarian Efforts in the Community, 1895–1897

Hard times had come to Central Florida following the Great Freeze. Hired hands and servants lost their jobs and collective belt-tightening became the order of the day. Since arriving in the Lake Charm area, the Meads had always practiced kindness and tried to help people less fortunate than themselves. Edith had early training as a nurse in the Brooklyn Homeopathic Hospital, and Theo had read and studied many medical books and had an excellent knowledge of plants as sources of herbal remedies. Now that economic depression gripped the area with the failure of citrus, there was a surge in demand in the local community for their medical care and advice. They willingly dispensed support, medication and simple advice about nutrition and care, particularly to the mothers of sick children, with Edith paying visits in the horse-drawn Murray phaeton.

The usual problems were with things like burns—Edith had plenty of experience with the severest kind—and the treatment of coughs and colds. Two extracts from Theo's letters to his parents show how busy they were, "Pretty much all the children in Oviedo have the whooping cough and E is in great demand to distribute the medicine Drosera—which we gather and make ourselves from the little sundew plants that grow in the border of Lake Charm," went one, and another, "All the hired people are kind in praise of E's Drosera. ... Mrs. Brock has whooping cough so badly that she almost suffocated and found great relief from the remedy."

Through this and other actions, the citizens of Oviedo held the Meads in high regard, although no doubt some still thought of them disparagingly as Northern Yankees. Their generosity and love of children were most evident around Christmas. Each year with little money, they made extra efforts to provide presents for all the neighborhood children, regardless of race, and helped decorate and dress the large communal Lake Charm tree with candles and presents. For Christmas 1895, Edith kept herself busy, largely doing things to keep from missing Dorothy, although Theo commented, "Seems as though finding fifty presents with only a couple of dollars to fall back on might keep anybody busy."

Christmas Day arrived and presents exchanged—Theo received various books: *Cuore* by Edmondo de Amicis from Edith, Huxley's *Science and Hebrew Tradition* from Mr. Edwards, and *The Prisoner of Zenda* by Anthony Hope from Mrs. Edwards, and three sets of pajamas Edith had made for him. Edith was presented with a beautiful writing set, a glass and silver inkstand in a stone stand bound in blue linen, and a dainty and pretty collapsible workbasket that lay flat for transportation.

Growing things to eat and selling the excess became even more necessary to meet day-to-day expenses. Theo planted large quantities of potatoes together with the usual beans, corn, tomatoes, and a cultivated bed near the packing house with 500 Gandy's Prize strawberries. Unfortunately growing more edible food had its drawbacks in the form of attracting wild hogs, the "razorbacks", that were apt to devastate crops with their voracious rooting and wallowing.

Although they were the property of the local farmers, many people were in the habit of shooting them as a nuisance. Florida law was a $1,000 fine or two years in the penitentiary for doing so, but realistically it was never enforced but diluted to allow the owner the value of the hogs if he could prove who killed them.

23.1: Florida razorbacks roamed freely in the Oviedo area and caused extensive damage to Theo's garden.

Theo had his own solution, poisoned cow feed, as he told his parents:

> The hog nuisance has become intolerable—they drive E off the lawn and are in the place all day and all night and have got so bold from our mild measures that they can't be driven out even temporarily, so we tried kindness once more and fed them with cow feed with appropriate flavorings: Paris green on Wednesday, pounded glass yesterday, corrosive sublimate today, and the menu for tomorrow is white arsenic.

Despite the poison, Mead's garden remained a magnet for the hogs even when he resorted to the use of strychnine, as Theo explained in a subsequent letter:

> Hogs ravaged the garden again last night, ate half the strychnine but Mrs. Swanson, who used to live in Cater packing house many years ago, says they fatten on it. The hog men cut all their fences as soon as the corn was nearly ripe. Mrs. S says the only way is to drive them into a pen or catch them with a pen trap and kill them with an axe. … I think I will use what barbed wire I have and run a fence from Ward's cottage to Lawton's fence corner by this house and try to keep our veg. garden from them as we depend so much on the vegetables. They rooted a lot of okra plants clean out of the ground last night besides other damage.

Theo's horticultural plants suffered too, and one morning he found his caladium beds "a tempest tossed waste of upturned stakes and labels and tubers." Hardly any part had escaped and there was nothing left to do but to rake out the tubers in a jumbled heap.

In September 1896, Edith was in Coalburg again, on this occasion to look after her mother and father and support Anne in her confinement for her third child. Because of the financial constraints, Edith's brother, William Seymour, did the inviting on behalf of his mother and sent $100 to pay for Edith's visit. Once there Edith found herself in the familiar position of being torn between duty to her family and pressure to stay there through Christmas, and support to Theo 800 miles away. She wrote to Theo:

> One reason why I dislike to stay through November is that my clothes are not warm enough for winter weather, and if I stay that long there is no telling how much longer I will or may have to stay. … There is almost nothing for me to do here during the next four weeks. They have good servants and my help is not needed in the house keeping at all. I wish, oh how I wish, we did not live so far away. It is hard to be dragged two ways

at once trying to be with you and yet having them want me here so much. What shall I do anyhow?

To pass away the time, Edith took up the bicycle as exercise, riding around the garden supported by friends on either side of the machine so she wouldn't fall off. Bicycling had become a craze by the 1890s and Edith's doctor thought it the perfect exercise for her. Several of her female friends who rode claimed that their headaches and other usual ailments had disappeared as if by magic. Theo was supportive of the idea and began to investigate the costs of purchasing a "wheel" for Edith when she returned to Florida. He had one little worry, however, regarding dress and wanted to know if women were now generally wearing bloomers to ride in, to which Edith replied, "By the way I may just mention here apropos of your letter that ladies do not go about in bloomers without skirts. They wear them generally under the skirts but have the latter short enough to escape the shoe tops."

23.2: *A photograph of the front of* Waitabit *cottage taken by Theo sometime in 1897, with Edith's bicycle parked against a nineteenth-century Chinese pottery drum stool.*

Back at Lake Charm, and with $55 of Christmas money to spend, Theo began investigating the bicycle question via nephew Archie Foss in New York. Archie identified an 1896 Eagle model for $50, slightly shopworn, otherwise new and guaranteed by the manufacturer that he had shipped to Oviedo in January 1897. Theo was convinced riding would make a marked difference to Edith's health. For her part, Edith had not lost her earlier enthusiasm for the exercise, saying that she was going to keep it up all summer by riding in the evenings after the heat of the day was over. She had a couple of falls, which made her cautious for a while, but kept the exercise going on a regular basis, practicing twice a day on a circuit around the house and "dodging through the bushes in fine style," according to Theo.

All his life Theo had used the flowers and plants he grew to beautify the world around him. He had richly decorated his study room at Ithaca with flowering and foliage plants and made sure that the fraternity parlors and meeting rooms at Cornell were adorned with whatever cut flowers or growing plants he could lay his hands on. At Dr. Foster's Lake Charm Chapel by the lake, on those Sundays when services took place, he performed a similar horticultural transformation to the interior. From his garden and greenhouse, he always had orchids for flowering interest and bundles of decorative ferns to provide the basis of an altar arrangement. In February, he might use pussy willow, peach blossoms, and Carolina jasmine, in March, sprays of the Cherokee rose, while the choice in summer might include the many varieties of jewel-leafed caladiums. His amaryllis were stunning centerpieces in March and April. At appropriate times of the year, he liked to use plants with some degree of religious symbolism—the pure white flowers of the Easter Lily or bracts of the peace lily (*Spathiphyllum*), or in late spring and summer, the purple passionflower. On Palm Sunday, what else but unusual and beautiful palm leaves from his extensive collection?

23.3: *The interior of the chapel on Lake Charm on Palm Sunday 1904, decorated by Theo with numerous palm fronds, cattleya orchids and, on the back wall, draperies of Spanish Moss spelling out the word "Hosanna!"*

CHAPTER 24

Financial Problems, 1897–1901

In 1897, the Mead family expenses were more than $1,000 a year and plant sales amounted to only $100, so they had to rely on continued parental support. Judicious disposal of the family assets and strenuous efforts by Theo and Edith to bring in small but welcome amounts of cash just about kept the financial boat afloat. He wrote to his parents:

> I have no animosity against my surroundings—would rather live here than in a cold climate but feel the imperative necessary of making a livelihood—i.e. enough to eat and wear. Positively the only money crop that is possible on our land seems to be caladiums—if I could grow $1,000 worth instead of $100 worth it might be worthwhile to wait for the groves.

Theo had been growing caladiums since his time in Eustis in 1886 and by the mid-1890s, although caladiums were in demand, they fetched only a few cents per tuber for the common sorts. "I put up all the caladiums I could spare today to fill the order—250 large and medium sized and 100 small for which I charge the very moderate price of $11.50," was his comment after one sale. He knew that better prices could be achieved by unusual varieties and set to work cross-breeding, using as his base a choice collection of 53 kinds of fancy caladium from E. O. Orpet. He described the leaves of these as "a culmination of silver and jewels and gorgeous stained glass," and added, "I notice quite a number of them are listed in this autumn's foreign catalogues at from $5 to $7.50 per bulb."

24.1: *Edith knitting outside* Waitabit, *circa 1900*.

Edith was playing her part too in supplementing the family income, running a sewing class on Saturdays and giving five paying music lessons a week to local children. Additional cash came from selling their excess milk—they were getting up to 17 quarts a day from their two cows—and they had permanent customers for butter over the summer, so long as they continued to buy ice. In late 1897, a fortuitous windfall of $886.87 arrived for Theo from the Alpha Delta Phi fraternity at Cornell in repayment, with interest, of the loan he had made twenty years ago for the construction of their first chapter house.

The winter of early 1899 was another cold one and the low temperatures cut back any new growth on the few trees that Theo had managed to save from the 1894–1895 freeze. On February 10, the mercury was 20 degrees at Lake Charm and it was snowing. He walked through the North Grove with Edith, who was amazed at the general destruction of the last two years new growth. The greenhouse remained at 40 degrees, but all the trays containing last summer's seedlings were frozen solid. In the outside garden, a few of his cauliflowers were slightly nipped and the beets had lost half their tops. Theo considered himself lucky that he hadn't yet planted out any of the tender vegetables, and summed up his feelings to his parents:

> As the settler said when he came home and found the Indians had burnt his cabin and slaughtered his wife and children—"This is a little bit too ridiculous." … I consider the orange grove as practically destroyed entirely—probably it would be as profitable to begin again with orange seeds as to bring them out of their present state. But of course we can't tell positively for a couple of months.

For the first time in his life, Theo was seriously worried about his financial future. The family had been using their capital for income for almost forty years, and there was only one property on Fiftieth Street in New York remaining. Available cash was tight. In July his parents leased the 800 acres they owned at the Lake of the Woods for turpentine production to J. W. Overstreet for $275. In Theo's view, his entire property would probably only fetch about $300 at auction, although it cost almost $30,000 initially. Over their years of ownership, he calculated he had paid out around $15,000 in labor and fertilizer and received $35,000 in income.

Aggravating him further was his mother's naive optimism over the future of orange growing in Central Florida and her belief that "everything was for the best." It drove Theo to distraction at times as his letters reveal, "I find Mamma's optimism rather depressing—it is so evidently without basis in any facts. The probabilities of the case all favor the view that profitable orange culture north of the 28th degree is as extinct as the mastodon." He concluded that as far as they were concerned their story was akin to the proverbial one of "from shirtsleeves to shirtsleeves are three generations," and had to be accepted with philosophy, if not with resignation.

The freezes of 1894–1895 and then 1899 had destroyed the dreams of even the most tenacious of orange growers, and Theo feared the worse when they received a letter in March 1899 from Alma Robie to Edith. In it she wrote that Harry had been very depressed all last summer—feeling that there was little or no chance of redeeming the grove even without this last calamity—and that they intended to quit Florida forever on May 1. She asked whether it might be possible for them to come over and see them before they left. Theo summarized this for his parents:

> After fourteen years of work, they go poorer in everything than when they came, except for the two boys and they might have had them somewhere else just as well. Harry will sell what tools, etc., he can and abandon the place entirely. Harry doesn't know what he will find to do at the North but will hunt up a job and Alma says she is resigned to the anticipation of living in two attic rooms in a city, though she regrets it for the boys who have been very happy in their country life and freedom.

The Meads were at the lowest point of their lives and virtually penniless. They calculated that the cost of going to Mount Dora to see the Robies represented five weeks of ice in the hottest weather, but Edith was of the opinion that even if it were ten weeks, she would rather go. They both agreed that if the worse came to the worse and they were fated to starve, they would rather have the end come one or two months earlier and have the memory of their visit.

They saw the Robies in April and there was an emotional farewell. Harry promised Theo that if he found anything good in the North, he would send for him immediately. In discussions about their future, Alma volunteered the thought that, with his encyclopedic knowledge and ability to make friends readily, Theo ought to be a teacher and run a boy's school. Theo was sad but philosophical about their leaving but admitted that he would miss the boys in particular. "If they had always lived close by, it would have at least doubled my enjoyment of life," he confessed to his parents.

In the dying embers of the nineteenth-century, the Age of Progress, Theo's father Samuel passed away, aged 76. He and Mary were staying at Eustis Bluff in their newly built house and enjoying the location as they usually did after the noise and pollution of New York City. A portion of a letter he wrote to Theo a few years earlier expressed his love of the region, "Mature trees a perfect glory of contrasting shades—couldn't be imitated … Air fresh, cool, delightful. Impossible to feel low spirited breathing such gaseous ambrosia."

His father took pleasure in rowing on Lake Eustis and frequently went by water to the Eustis post-office and other stores, a journey that with a strong headwind could take up to an hour. But his boat was on its last legs, leaking water where the canvas had separated from the wood, and in urgent need of repairs. On the morning of August 21, 1899, in apparently the best of health and good spirits, he breakfasted as usual then made his way down the inclined slope of the bluff to the boathouse to begin repairs. At one p.m., when he had not returned for his dinner, Mary went looking for him and found him sitting on the steps inside the boathouse, having been dead apparently for some hours. Mary summoned Theo, who arrived on the train the next morning. Together with a local doctor, they judged that the cause of death was a clot in the brain (cerebral hemorrhage), which seemed to be a reasonable and probable explanation of the facts.*

---

*Many years later, Theo discovered the real cause of his father's death. His father had a terminal cancer that he kept from the rest of the family, and chose to end his life on his terms via a self-administered overdose of chloroform (Information contained in letter TLM to Willis family, July 24, 1934, Willis family collection).

Mary wanted Samuel's body transported back to Brooklyn and buried in the family vault in Green-Wood Cemetery. Theo agreed to help but pointed out that it was not possible under Florida law to move the casket until the expiration of a full year, and then someone must accompany it on the train with a doctor's certificate. This gave them time to consider the future of the Eustis property. Both agreed that for the moment, pending its sale, they should board the house up and employ a local caretaker to perform essential maintenance. Mary decided she would return to New York City where she had friends and relatives to look after her, and eventually moved into a house in Brooklyn at 438 Classon Avenue, owned and lived in by her widowed brother Francis, ten years her junior, his daughter Jennie and sister-in-law Cornelia Suydam.

Theo's father had always been a positive supporter and sounding board for ideas and things Theo had wanted to try. He remained neutral in matters of religion and was a significant stabilizing force between Theo and his mother, staying on the sidelines in the many religious debates they had. After his death, Theo continued to write to his mother but greatly missed the practical and business advice frequently contained in his father's letters. He needed another trusted and friendly soul to talk to, especially with respect to his financial position, and turned to his brother-in-law, Will Edwards.

Over the Christmas period and into the start of 1900, Theo had time to take stock of his financial situation and wrote to Will at Coalburg, unburdening himself with his troubles and asking for advice. Will's kind reply came on January 20 where he expressed the view that he did not believe that Theo was "personally to blame for the disasters that Nature seems to have poured upon the orange growers of Florida." He advised trying to make a go of it with caladiums and added it distressed him greatly to think how little they lived on. He ended, "All I have is for those I love and my only grief is that I have not been able to do more. But keep your hearts brave, both of you, and remember that so long as I live you shall never want."

With his father gone, this was just the caring boost Theo needed. Further good news reached him later that month with the eventual sale of the last of the family's Manhattan properties on West Fiftieth Street, netting Theo and his mother $5,000 each. Both agreed that this money should be invested to provide a regular income. Theo once again sought advice from Will, who was then the manager of The Vespertine Oil & Gas Company of Charleston, West Virginia, drilling for oil and gas near St Marys. He suggested that $5,000 could buy a thirty-second interest royalty in one of the drilling activities and this should yield not less than $50 a month, figured as ten percent interest on the original investment plus $100 a year credit on the principal. Theo signed agreements to this effect and replied gratefully to Will that it was "brotherly of you to let me in on the ground floor."

24.2: *Mr. W. H. Edwards, Theo's father-in-law, aged 80, sitting in his favorite cane chair outside Bellefleur at Coalburg in 1901.*

A letter arrived in early May 1901 to brighten up Theo even further. It was an invitation from Martin McVoy, the Chairman of the Executive Committee of the Cornell Chapter of Alpha Delta Phi, for Theo to attend a grand reunion at the Alpha Delta Phi Club, 35 West Thirty-third Street in New York on Saturday, May 18, to help promote the new chapter house. McVoy told him that many of Theo's old Cornell chums, such as Tut Morris and Will Edwards, were planning on attending and at any get together of the fraternity things were not the same "without your face being present and your voice singing 'Taranty my son'." As if this wasn't enough inducement, he added that since Theo had been so instrumental in the foundation of their first chapter house, his presence would be essential and so all expenses would be paid as part of the invitation.

In 1901, there was no money in the kitty for Edith's almost annual summer return to Coalburg and Theo didn't want to ask Will, the usual provider of funds. All this changed however on August 15 when Edith's mother suddenly died and Will wired her money for the fare to come and join the grieving family and help look after her frail father.

Some further easement to the family budget occurred in October 1901 when the Lake of the Woods land, bought as an investment by his father and previously leased for turpentine production, was sold to S. W. Eldredge of Apopka for $800, the same amount as paid for it twenty years ago in 1881.

CHAPTER 25

Acceptance of Faith, 1903

Since the passing of Dorothy, Theo had been deep in thought about his religious beliefs. For the last 30 years, he had been a confirmed agnostic, but some of the church services following Dorothy's death had moved him. A chance encounter in April 1895 with a young boy taking communion at the Episcopal Church in Winter Park affected him deeply. He tried to explain the incident to his parents:

> The sight of a friendly boyish face and the sound of his voice seemed to relieve some little mental tension and made me feel happy and glad and in sympathy with my surroundings—a frame of mind which I attributed largely to relief from the tension caused by unsatisfied parental instincts and supposed it would soon pass away but of course I enjoyed it while it lasted. … it occurred to me that this was what people call the Peace of God which passeth understanding and might last always if I could keep attuned to whatever love and right feeling I could perceive or feel, whether we choose to call it God or simply love.

From his Ithaca days, Theo had attended church services and heard many preachers. He recalled what he had heard Phillips Brooks say in one of his sermons, "It is not whether you love God, but whether God loves you." He was also greatly impressed after reading a series of essays and notes by George Romanes entitled *Thoughts on Religion*, published in 1895. Romanes, like him, became an agnostic because of Darwin's work on evolution but in later life moved to accept a Christian God through the vehicle of scientific reason and logical analysis.

Theo believed he had felt the love of God in his heart in that simple church in Winter Park, but he realized that this shift after so much time would need some explaining to his parents. A series of letters followed and in one he wrote, "In following out Romanes line of thought it seems impossible for me to do otherwise than agree with him, that it is reasonable to be a Christian believer." Theo went on in subsequent letters to develop and express his views that religion could be disastrous when not controlled by reason and science, that religion was solely a personal affair, and that it was unreasonable to expect any two people to hold the same view of the Almighty.

His mother was initially delighted with Theo's change of heart and wasted no time in enthusiastically telling friends and family about the conversion. Letters came from family members congratulating him on becoming a fully-fledged Christian that had Theo irritated, then as they continued to arrive, hopping mad:

> I wish if possible to explain to Mamma some reasons why I do not wish her to execute a religious war dance over me—or in any other way to abuse the confidence of an uncalculating moment of enthusiasm. Please remember at any rate that I am already past the meridian of life, and quite competent to decide for myself what publicity should or should not be given to my opinions and emotions. You see that every proposition of yours for publishing abroad my private confidences acts on me as a red rag to a bull, and I propose to make my protest just as forcible as my command of the English language will allow. I don't see why you shouldn't be satisfied with saying that I thought of becoming an Episcopal Church member. Surely people are capable of drawing sufficient inferences on all points.

But his mother was not satisfied and the letter exchanges continued through the rest of 1895, becoming increasingly rankled. "Mamma says she hopes I will be a 'humble Bible Christian'," wrote Theo, "If you mean by this that I shall make a fetish of the Bible and bow in idolatry before the 'letter which killeth' instead of trying to appropriate the 'spirit which maketh alive' I am afraid you will be much disappointed."

Theo believed that the stories in the Bible were largely allegorical and that Jesus would not have performed miracles "for the sake of surprising a gaping crowd of peasants in Judea," pointing out not that he couldn't, which was absurd, but that he just wouldn't. Original sin, forced devotion and strict obedience of the Sabbath were further religious dogmas dear to his mother that did not make sense to Theo. He tried to explain to his mother his attraction to Episcopalianism:

> In the Episcopal church you have to assent to practically nothing beyond the Apostles' Creed and are not tied down to any one interpretation of that, either. You can believe as many dogmas in addition as you like and the majority of Episcopalians do but they needn't unless they choose as far as church membership is concerned.

But to his mother this was heresy and she stepped up the rhetoric as if Theo had just confessed to a fetish of worshipping the Devil. From nagging him about non-observance of the Sabbath in her letters, she moved on to hinting at a darker implication—that some of Theo's misfortunes might be a punishment from the Lord, "So sorry you travel on Sunday and do not worship as a rule with His people on the Lord's day. I believe it displeases the Lord and He is showing you marks of His disapprobation. What does the death of the cow mean? What has the Lord against us?"

Theo could accept a loving God but not a vengeful one, nor a vain one who needed praising all the time. For him if his cow died it was not a signal of God's disapproval but just bad luck; if he worked on a Sunday to provide for his family or wrote a letter it should not matter to a loving God who did not demand strict obedience and constant praise. Once again he tried the voice of reason, arguing that as the proverb goes "what's one man's meat is another man's poison," particularly when applied to religious matters. He pleaded for a compromise, writing:

> As Mamma, all her life has been considering chiefly the things she has been told—and the telling is all through human sources and none of it at first hand—and I have chiefly studied what we can know of God's ways and habits, it is natural that there should be no end of points of disagreement. It would be better maybe for both of us to lay all the stress on the points where we <u>agree</u> rather than on the points where we disagree.

But his mother was impervious to reason when it came to religious interpretations other than her own and ignored the olive branch. In response to Theo's diplomatic communication, she delivered one last barb aimed at the heart of his life's work—successful experiments in horticulture. These counted for little in her view while he remained outside her definition of a "humble Bible Christian," and she had no hesitation in telling him so. He cut out the offending letter extract and sent it back to her with the comment, "I refer you to St. Paul who remarks 'Be ye content with such things as ye have'."

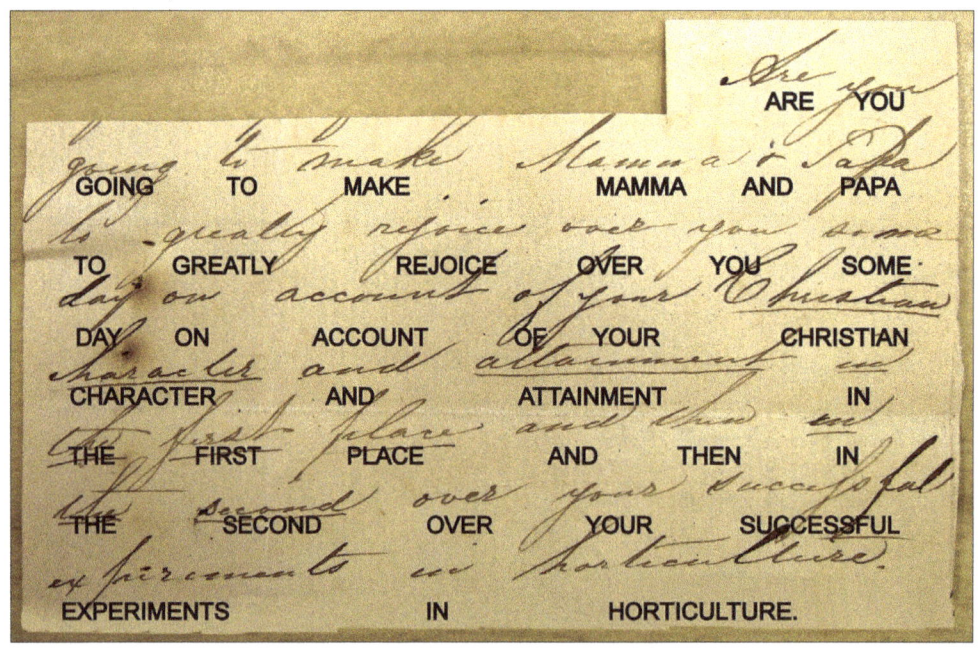

25.1: The offending part of Mary Mead's letter to Theo.

190   Orchids and Butterflies

Neither side had moved an inch in the debate that had been going on all their lives but had intensified over the past year. Theo decided to stop trying to explain his beliefs, with one last closing remark on the subject:

> If Mamma had never said or written a word to me on the subject of religion—say for the last 30 years—but had simply lived her life of love and self-denial and trust in God without thrusting her own private "leadings of the Spirit" upon people whom the Spirit led differently, it would have been inexpressibly easier for me to accept the Christian truth.

In 1903, Theo's religious affiliation to the Episcopal faith, forged out of the aftermath of Dorothy's death, the sermons of Phillips Brooks and the writings of George Romanes, took its final step with Theo's decision to be confirmed. In April, he wrote to Bishop Gray at the Cathedral Church of St Luke in Orlando, who replied that he would be delighted to perform the ceremony and suggested Ascension Day, May 21, as a good day, it being special to him in marking the forty-third anniversary of his advancement into the priesthood.

On May 21 at 11 p.m., the service took place, with Rev. Frankel reading the lessons, Rev. Perdue the prayer, Rev. Rickert preaching, and Dean Spencer acting as server. Bishop Gray confirmed six persons—three from Winter Park and three from Oviedo. Theo's mother was keen to hear of his experiences, writing, "I am waiting to hear you recount the experience of the 21st. I hope it was a very happy one. Those are lovely verses by Phillips Brooks and I am so glad they express your inmost soul's experience."

A congratulatory letter from Julia Inness evoked a response from Theo that explained his acceptance even as a committed Darwinian:

> When the immortal and the "Kingdom of Heaven" are seen to be simply the crown and capstone of the whole system of evolution—the end and aim of the whole process from the nebula and the primordial slime up to the human mind and soul and we know not how far upward and beyond—the

difficulties in the way all seem to vanish, and we can repent to realize that though surely descended from the slime we are nevertheless, as you say, the Children of God and every right to trust in His loving kindness.

25.2: Theo was confirmed by Bishop Gray as a member of the Episcopal Church at the Cathedral Church of St Luke in Orlando.

Edith had closely followed the religious discussions between Theo and his mother. She was pleased and sympathetic with the outcome, although not very demonstrative, which Theo described as "just right."

25.3: Theo photographed in September 1903 shortly after his confirmation into the Episcopalian faith.

CHAPTER 26

Medical Issues & Loss of Family Members, 1902–1916

Edith was still prone to frequent and acute headaches and a further complication arose through the late 1890s in the slow development of an area of soreness on the back of her right hand, which stubbornly wouldn't heal despite ointment and bandaging. It caused pain and irritation and began to interfere with her piano playing and household chores. By the middle of 1902, soreness and a dark crust had developed and it was decided that Edith should go north again to Coalburg and seek treatment with a specialist, Dr. Schoolfield, in Charleston. For Theo, the worry was that the affliction was cancerous. At Charleston in October, Schoolfield and a number of other experts examined Edith's hand and X-rayed it. Edith wrote to Theo that "they found nothing beneath the surface and pronounced the trouble eczematous … it is nothing dangerous or likely to become so. Cancer is quite out of the question in any event." Her hand was on the mend and Dr. Schoolfield saw no reason why Edith should not return to Florida. But there was another reason for a little longer stay in Coalburg with the news that at the ripe old age of 48, brother William Seymour was to be married in London in July and would return to Coalburg in late October after a honeymoon in Norway. After this reunion, Edith returned to Lake Charm.

In 1907, her hand had started to trouble her again. News from Coalburg was that her father was becoming increasingly frail, so once again a summer visit to Coalburg was agreed upon followed by a period at Clifton Springs to build up her strength and get medical treatment for her hand. Theo remained at Lake Charm, working hard in his vegetable garden where there was the promise of a bumper crop. But as part of this general decline in family well-being, even Theo came down with a heavy cold. His mother was initially sympathetic, "So sorry to hear of your severe cold," she wrote but couldn't help following it up with, "I believe it is sent in mercy to show you the folly of being too intent on worldly gains."

When his mother heard of Edith's potentially long stay at Clifton Springs over the summer and early fall of 1907, she offered to come and help look after Theo and keep him company at Lake Charm. Theo accepted the invitation. The community around the lake had shrunk significantly since the Great Freeze; almost all cultured Northerners had left and Theo was feeling starved of intellectual conversation. He told his mother, "Even now Mr. Hurland is the only person here for the winter who is capable of <u>thinking</u>—to any purpose that is." By August, Edith was with her family in Coalburg and a few weeks later Theo's mother arrived at Lake Charm.

Edith found her father frail and she was shocked at how weak he had become, writing to Theo, "Father was up at ten and afterwards I took my work and went upstairs to sit with him. Just now the mail came, and I was shocked to see that he is so weak he can scarcely turn his newspaper over. He looked at it a few minutes, and then said he was tired and must lie down again." There was a letter in the mail from a butterfly collector wishing an audience with him, but both Edith and Anne thought this out of the question and sent back a letter apologizing and explaining he was too feeble to talk.

Edith arrived at Clifton Springs in late August to begin a course of treatment and await the arrival of Dr. Tinker, a skin specialist from Ithaca, to look at the hand. News of his mother's arrival at Lake Charm had Edith fretting, "It was not a well-advised move on her part, when you have so little time to visit with her and can really do so little for her entertainment. I hope on Sundays at least she will see the need of your resting and not insisting on church at any cost to you."

Medical care at Clifton Springs for the hand involved a course of X-ray exposure every other day and dressing with ointment. The bandaging made it hard to hold a pen to write to Theo, so she had a go with her left hand. New skin was growing again, but a skin graft was judged necessary, which would incapacitate her hand entirely for a couple of weeks. It was the end of October before she regained full use of it and could return home.

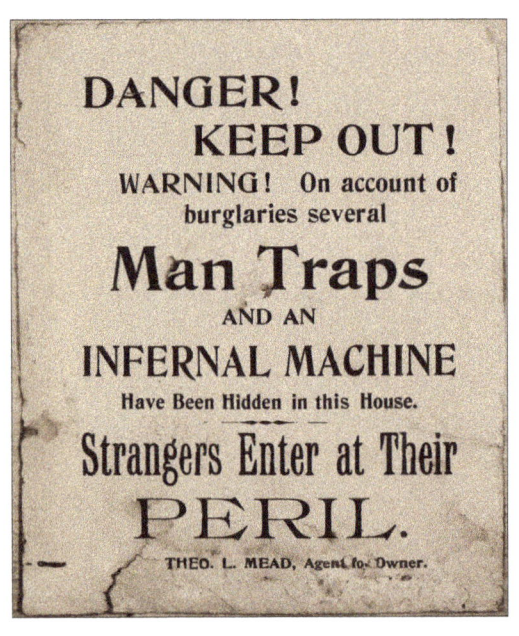

26.1: *The house of Theo's parents at Eustis Bluff had been left empty pending a sale. Theo had this poster printed and attached to the door as a warning sign.*

Back at Lake Charm, on one occasion Theo went with his mother to Eustis Bluff, where Mary and Samuel's original home had been boarded up and a caretaker manager employed to keep an eye on things. Unfortunately, they found the place had been broken into, oranges stolen and the gardens nearly ruined. Thieves had forced the boards off the front entrance and broken a pane of glass to gain entry. Theo decided it was time for action and printed a poster that he pasted to the door, warning potential intruders of the presence of mantraps and, hidden inside the house, an "Infernal Machine."

26.2: *Mary Mead photographed in 1908 on the steps of the New York Bible School at Lexington Avenue and Forty-ninth Street in New York.*

Before the end of the year, Mary Mead had returned to New York to the house in Brooklyn. Always on the go despite her advancing years, she was a regular churchgoer at the Dutch Reformed Church on Fifth Avenue and Forty-eighth Street, where she had been a member for almost seventy years, and also an occasional contributor to the New York Herald *Sunshine Club*. Both of Theo's parents had been ardent supporters of life-long learning through the Chautauqua movement and correspondence courses, and for Mary one of her key objectives was to master the Greek language.

In 1908, she enrolled at the New York Bible Training School on Lexington Avenue, attending classes in homiletics, church history, the prophets, pastoral theology and archeology, and making a specialty of the Greek language. Each morning, six days a week, she would rise early for her commute into Manhattan, traveling on the Gates Avenue car and then a Third Avenue car from the Bridge, arriving before ten o'clock to join the rest of the young students at morning prayer. As one of the oldest students and commuters between Brooklyn and New York, she was popular with the young men and women that the school trained to become Bible teachers and mission workers, who described her as being "eighty-five years young." In between lessons, she would move to a reception room and perform a series of gymnastics and breathing exercises.

When asked the purpose in studying Greek, her reply was so she could read the New Testament in its original form and "enjoy a deeper communion with my Lord." This remarkable example of life-long learning appeared as a story in the *New York Herald* of 1909 under the headline, "Woman a Greek Student When 85 Years of Age—Mrs. Mary C. Mead travels from Brooklyn to Manhattan daily to attend classes in Bible training school."

In April 1909, Theo's father-in-law and mentor to his butterfly achievements, William Henry Edwards, died at Coalburg aged 88, leaving no will, no property, two honor debts of $4,000 and $10,000 and a study described by his son, Will, as a "mass of indiscriminate papers."

Will Edwards had made a lot of money out of the coal, oil and gas business in West Virginia, and always looked after the family financially where he could, funding many of Edith's trips to Coalburg. His sister Anne had always painted from the early days of drawing and hand coloring caterpillars and chrysalis for her father, earning the occasional assignments for special events and gifts but not succeeding in making many sales. The picture dealers generally priced her pictures at $20 to $25 framed, if selling on commission. Theo had sometimes used Anne's talents, on one occasion in 1893 to paint a watercolor of his Cherokee rose hedge at Lake Charm as a wedding present to the bride of one of Theo's old Cornell fraternity brothers, Clarence Esty. After the death of his parents, Will had amassed enough money through his own efforts to treat Anne to a trip to Italy to see the classical works of art and asked Edith to accompany her over the winter of 1910.

26.3: The Cherokee rose growing at the edge of the Mead property and Lake Charm was a favorite flower for chapel decoration.

While Edith was in Italy, Mary Mead once again joined her son at Lake Charm to keep him company and was with him over the 1910 Christmas period. Theo kept himself busy tending to the vegetables and his newly planted amaryllis bulbs, while Mary did the household chores and wrote letters to friends and relatives, sitting outside in the shade of the garden.

Improving the mind and keeping fit and active with a healthy diet remained essential aspects of Mary's life, even now that she was in her late 80s. She was still actively learning, and the newly rebuilt Battle Creek Sanitarium in Michigan provided a structured environment for all these things—a strict diet along with vigorous exercise, plenty of fresh air and rest. She checked in for an extended stay in July 1911.

26.4: *Theo's mother writing letters in the garden at Lake Charm in 1910.*

A January 1914 letter from Brooklyn to Theo from his mother, now aged 91, presaged the events that would unfold later that year, as she wrote:

> My darling son Theodore, I do not feel very well. My feet are swelling more and I have a sense of fullness about the heart and stomach, and I breathe quickly, a good part of the time. Last night I had little or no sleep and this is habitual with me. I do not wish to complain at all, but it seems to me that this state of my bodily condition cannot last long.

Over the summer, she moved to the sanitarium at Battle Creek again where staff knew her from previous visits, dying there on November 9, 1914. The official death certificate recorded the causes of death as senility, myocardial degeneration, and arteriosclerosis, common euphemisms at the time for "old age." Theo went north to transport the casket back to New York and attend the funeral, after which Mary Mead's body joined her husband and eldest son in the family grave at Green-Wood Cemetery, Brooklyn, at Lot 738, section 104. Mary Mead left no will, and the Kings County Surrogate's Court recorded the remnants of the Mead/Luqueer family fortune and appointed Theo as administrator. The estate amounted to about one thousand dollars in personal assets and property worth an estimated thirty-five hundred dollars.

At Lake Charm, the routine of Theo reading aloud to Edith after supper continued, and around 1915 they bought a Victrola Talking Machine and added variety to the evening with music. They soon amassed a considerable record collection but their music tastes were very different. Theo, who had a clear, strong voice and always loved singing from his fraternity days, preferred popular songs and hymns sung by tenors, while Edith was drawn to the music of the great classical composers such as Beethoven, Mozart, and Brahms. They came together musically on Sunday evenings which were always hymn nights. He described the arrangement reached to accommodate their different musical tastes to a friend:

> My wife and I have a funny arrangement about getting new records—I used to stay up till 11 or 12 o'clock reading the war news and my wife said she'd make a bargain with me—for every half hour after 10 that I stayed up I must pay her a nickel towards getting classical records that she likes, and for every half hour

that I went to bed before 10 she'd pay me a nickel to get frivolous ones, like Tipperary or Wynken, or any kind I wanted. So I immediately reformed and so far have earned about $30—she says it's rather expensive but worth the money to have the house quiet and lights out after 9. We have a nice collection of records now—about 230 pieces of just the kinds of music we like.

Their 1915 Christmas was tempered by the news that on Sunday, December 26, 1915, Edith's brother, and Theo's close friend from his Cornell days, William Seymour Edwards (Will), aged 59, had died at the John Hopkins Hospital in Baltimore after a long illness.

26.5: William Seymour Edwards (Will), Edith's brother and Theo's Cornell Alpha Delta Phi fraternity friend.

The last few years of Will's life had been difficult. New well drilling stopped because of the war and a depressed oil price, and the illness that struck him had affected his ability to manage the oil businesses. There was a general feeling in the family that financial anxieties may have hastened his death. Will was always generous to everybody connected with him, and he left $100 a month to Edith and the bulk of the rest of his assets to his wife Hope and family.

Edith remained active in the local community, providing piano lessons and tending the local sick children. In the spring of 1906, she had been one of seven women who decided to form the Oviedo Magazine Club, whose purpose was to disperse worthy reading materials and improve their community. They established a library of over 400 books open to the general public one afternoon a week in the real estate offices of O. P. Swope, and hired help to pen up loose pigs, lay board and clay sidewalks, plant trees and clean up the cemetery. In 1916, they joined the Florida Federation of Women's Clubs and changed their name to The Oviedo Woman's Club. Edith was their first president, and for the club flower, they chose the Cherokee Rose that grew in abundance along the boundaries of the Mead property and Lake Charm.

In August 1916, a severe nose and ear infection laid both of them low. It started with a severe nasal cold that Theo caught and ended up passing on to Edith, infecting her ears and requiring surgery, as Theo explained:

> One of the pestiferous little crawling gnats we have here flew up my nose and being loaded with a select assortment of microbes, gave me the worse cold I ever had—I got over it in three weeks but my wife caught the infection and it went to her ear and we had a dreadful time for three months culminating in a "radical mastoid operation" which was delayed because it was feared that she could not survive the anesthetic. But she did remarkably well and in two weeks more (eight weeks from operation) will be able to dispense with further medical attendance. Meanwhile I got to my garden about once a week and for maybe 20 minutes at a time.

26.6: *Edith relaxes with the newspaper outside* Waitabit *in this photograph taken by Theo in 1915.*

Part Seven

The Master of Horticulture

CHAPTER 27

Orchid Breeding, 1891–1925

It had been many years since the Great Freeze of 1894–1895 when Theo had been forced to refocus his horticultural efforts into growing more food to eat and selling the excess. Now, for greater income, he decided to try hybridizing plants with significant financial potential. Orchids were the obvious choice for big money if he could market the results and afford to wait for the growth of mature, flowering plants. Word from the London orchid market was that unusual orchid hybrids were in high demand, some selling for up to $1,200 per plant. Theo had made his first orchid cross between *Cattleya schilleriana* and *Laelia purpurata* in July 1891 and by 1896 he had 40 or 50 blooms of various sorts in his greenhouse. But until he could build up a large enough collection of rare hybrids, he would have to rely for income on occasional sales of caladiums, palms, and other plants, supplemented by selling vegetables, eggs and milk.

A favorite source for new orchid material was the New York auction rooms where he asked his mother to bid on his behalf, selecting by choice the bargain boxes of miscellaneous items and leftovers. He later stated:

> Priceless novelties brought their thousands, but a dollar or two bought the promise and potency of less rare though equal beauty, and the mixed boxes of odds and ends and left overs among the orchids newly torn from their native jungles were to a Floridian like myself the most fascinating of all. Two or three years of loving care often brought gorgeous blooms from dry unpromising bundles of greenish sticks with hardly a leaf to recommend them.

Throughout the rest of the 1890s, his collection increased dramatically. He moved his tender palms out of the greenhouse to a sheltered place outside to give more room for orchid experimentation, and installed hot water pipes to protect the plants against winter freezes. Even with Theo's outstanding technical expertise, orchid hybridizing was a complicated business. Via correspondence with some of the leading orchid growers in the United States, he acquired viable ripe pods and pollen to use in his own hybridization experiments. For his own in-house efforts, he had only a few days after the flowers opened to complete the task before the blooms faded.

During 1893 and 1894, he had recorded 242 crosses from *Laelia* and *Cattleya* flowers alone, with *C. intermedia* giving perfect pods with good seed from *L. purpurata, L. granis,* and *C. Mossiae.* The bigeneric cross between *Epidendrum fragrans* and *L. flava* was growing nicely but the pod from *E. fragrans* x *C. skinneri* had yet to ripen. Overall *Cattleya bowringiana* gave him more hybrids brought to blooming size than any other species, and he was most pleased with Cattleya crosses involving *C. amethystoglossa, C. skinneri, C. maxima, C. forbesii, C. velutina, C. granulosa, C. schroederae, C. trianae, C. warscewiczii, C. dowiana* and *C. lawrenceana.* However, *Cattleya skinneri* proved to be a difficult species to cross, many times producing only a few seedpods that were very slow to ripen.

The leading horticultural question in orchid culture at the time was related to seed germination and why it was such a hit-or-miss affair. The ripe pods, whether his or provided by another orchid grower, provided the means to try and answer this question, and he set to work designing a series of experiments to improve the success rate of germination.

In the first experiment, he took an incubator consisting of a half-gallon candy jar containing fern root on which he sowed the seeds and set it up near an artesian well. Warm water from the well was directed to flood the outside of the jar and feed a small vacuum pump that sucked filtered air through a series of glass tubes attached to the top of the jar. With this arrangement, germination was improved but growth was inconsistent and slow. He reasoned that healthy orchid growth must require significant airflow, so he needed either an open external system in a breezy location or a larger closed one.

His second approach was to build an orchid eyrie as a small platform high in a live oak tree, with the seedbed kept moist by irrigation from a water ram with fine cyclone nozzles. However, although germination took place, midge larvae in the compost attacked and devoured many of the tiny plants, and it was difficult to keep the compost in good condition during the hot days of summer. He had some germination with this method, planted the successful seedlings out in trays, and transferred them to the heated greenhouse. In mid-1896, he had over a thousand mixed seedlings and set himself the target of growing a thousand to maturity each year. In 1897, he purchased a small water-driven blower for the greenhouse to create a gentle breeze over the seedlings.

His best results finally came when, with Bernard's discovery in mind,[*] he came down from the trees and tried his third approach of a closed system. He designed a glass-covered case about thirty feet long, into which he placed seeded pillows of cheesecloth. These contained ground-up decayed oak leaves and sphagnum, first sterilized by heat and then mixed with symbiotic fungi obtained from the little orchid plants that were a by-product of his experiments. Currents of moist filtered air fed the case from a second vessel containing live sphagnum and a lamp stove to keep the air warm on cold nights.

Theo kept detailed and accurate hybridization records, not only of orchids but also of other plants. His various notebooks and an index card system ran to

---

[*] An orchid seed has an embryo but no food store and cannot develop without an external food supply. This nutrient source was unknown until the discovery by Bernard in 1899 that germinating seeds always were associated with fungi as a symbiotic, food-supplying partner.

almost 300 pages and documented several thousand orchid crosses in the period 1891 to 1925. To complete the documentation process, he photographed many of the orchids at flowering time and meticulously hand-colored the resulting prints. He regarded his new creations as his children, and showed infinite patience, first in their germination and then in their flowering. He was perhaps most satisfied with the cross between *C. Lueddemanniana* x *L. Tenebrosa*, which he named using his system, Lc. Ludbrosa. This germinated in 1897 and flowered in 1914—seventeen years from seed to first flower.*

27.1: *Successful pillow germination of rooted leafy plants, averaging half an inch across, of* Cattleya trianae x Laelia flava *planted June 16, 1904, using seeds from E. O. Orpet, and photographed by Mead in 1905.*

---

* In 2014, this hybrid was reclassified as *Cattleya* Ludbrosa, and officially attributed to T. L. Mead.

27.2: Mead's orchid album resides in the Rollins College archives, Winter Park.

Theo was a skilled and dedicated orchid hybridizer, communicating with leading American figures of the time and becoming well-known in America for his unique collection. His main pollen sources and fellow hybridizers were Edwin Orpet and Oakes Ames, and one of his customers for orchids was the Royal Palm Nurseries operated by Egbert Reasoner in Manatee, Florida. He struck up a friendship by letter with Robert Allen Rolfe, the editor and founder of the London-based, *The Orchid Review*, who encouraged him to submit articles on orchids to the journal. Also, he wrote extensively on the subject of orchids and his experiences for other journals and horticultural proceedings.

In 1909, Rolfe and Hurst's *The Orchid Stud Book* was published, and this allowed Theo to separate out those crosses (hybrids) of his new to the world and those that were remakes of the previous efforts of European and American growers. He concluded that twenty-six species of *Cattleya*, twenty-nine of Laeliocattleya and five of Epicattleya were unique to him. In recognition of his efforts, the International Orchid Register named two of his *Cattleyas* in 1904 in his honor—*C x Meadii (bowringiana x forbesii)* for the man, and *C x Oviedo (schilleriana x maxima)* for his residence.

Bernard's research work in 1899 had concluded that germinating seeds always contained fungi as a symbiotic partner but the critical step, yet to come, was an understanding of the mechanism of this relationship and thus the development of an alternative non-symbiotic method of culture. This would have to wait until the early 1920s, and Mead would play a critical part in the discovery.

When Lewis Knudson at Cornell University started work in 1917 to understand the role that fungus played in symbiotically supporting orchid seed germination, it was natural for him to turn to Theo, who was by that time an orchid hybridizer of international repute. Knudson's objective was to find an alternative germination method that did not need the presence of fungus and gave controlled, reproducible and high germination rates. But he quickly found he could not commence this work without viable ripe orchid seed, and the few growers he contacted refused or were reluctant to supply him except at exorbitant prices. When he wrote to Theo, he found a ready and willing co-operative soul who not only was prepared to give him all the viable seed he wanted, but also joined enthusiastically in all the experiments and grew many of the seedlings on to mature plants.

Knudson and Mead exchanged dozens of letters over the two or more years the work took, with Theo always suggesting slight modifications or offering his interpretation of the results they observed. He proposed compositional changes to the media used for germination and actively monitored and described the degree of development of the seedlings within the many vials of tiny germinating seeds Knudson sent to him. Finally, success came with an agar solution containing a particular selection of mineral salts and nutrient sugars.

It is highly unlikely Knudson would have been successful without Theo's 30-year experience of orchid culture and the seeds he provided. His intellectual involvement in the work was also considerable and should have earned him co-author status on the paper, published in 1922, showing successful germination without the presence of the fungus. Unfortunately, Knudson did not see the need for this and settled for a couple of brief mentions of Mead's work in the paper and an acknowledgment of Mead's assistance at the end, retaining any fame that might come from this discovery for himself.

As a result, Knudson is the name remembered and Knudson C medium is commonly used today to germinate orchid seed of some species, and is still the only way new hybrids can be efficiently produced. With non-symbiotic germination technology, thousands of seedlings could be raised to maturity from a single seed-capsule to produce large quantities of plants economically. This scientific breakthrough made commercial orchid growing possible. Together with tissue culture, it launched the orchid industry we know today and ensured orchid affordability to the public.

27.3: Theo never lost his passion for the orchid, and continued to hybridize the genera into his eighties. In this photograph, taken May 1927, he sits in his greenhouse with some of his orchid creations.

CHAPTER 28

Citrus Freeze Protection, 1899–1905

By the late 1890s, some citrus growers in Central Florida had started to replant, but their chief worry was finding a way to economically protect the trees and citrus fruit from injury when "a cold wave a thousand miles wide and ten miles deep sweeps down on us from the far Dakotas." With his scientific and engineering background, this was just the sort of challenging question that Theo relished.

He knew there were two accepted protection methods, warming the air by burning fires and providing a shed or shelter for the trees, but both had had varying results in practice. Orchard heating worked on calm nights without rain or sleet, but sheds or shelters were vital when wind chill was a factor. He concluded that a combination of shedding with heating seemed to be the best answer. But, he wondered, why supply heat in Florida by lighting fires and burning fuel when there was an unlimited source of warm artesian water below ground, and with pumping equipment the means of bringing it to the surface? However, several calculations later he realized that it was not feasible to supply the vast quantity of underground water needed to warm the air adequately during a windy, severe freeze. He needed a new way of looking at the problem.

His prior experience with citrus had shown him that fruit chilled to the point of freezing at 32 degrees, and even below this to 28 degrees, did recover and was saleable. Below 28 degrees internal ice crystals formed turning the fruit to mush. He also knew from his scientific background that the process of water freezing takes place at 32 degrees and the temperature does not fall further while freezing is taking place so long as there is still liquid water present. With this arrangement, he reasoned, the fruit should remain undamaged. This led him to postulate a frost-proof orange grove with constant overhead warm-water irrigation that would allow the citrus to survive encased in their own ice cocoons, no matter how low the mercury fell. The only downside he could think of was the amount of water applied; too little and the ice cocoons would be dry and cool down further, too much and a large deal of ice might form, risking breaking the tree's branches.

He suggested this irrigation system would work best if the grove were contained within a shed to contain the warmth of the artesian water and limit wind-chill. He worked out the approximate cost for a wood and cloth-covered shed and wrote up a detailed account of his thinking. Unfortunately, the cost of the shed for an acre of citrus was $1,200, well outside his existing budget. However, Theo had a chance to share this vision with Edith's old school friend, Julia Inness, who visited Lake Charm in February 1900. Julia agreed to lend him the money, so Theo set about constructing the shed over the summer of 1900 and planted 475 five-year-old orange, tangerine and grapefruit trees in it in December. He wrote enthusiastically to a friend that things were "coming his way again, a little," and that he hoped he would be able to defy the frost with the shed and share the prosperity of people living 100 miles south. These growers, he declared, since the last crop of oranges "have been simply wallowing in money—getting in some cases as much as $4,000 an acre for a single crop." To hedge his bets, between the citrus rows, Theo planted tender vegetable crops such as eggplant and cucumber.

28.1: Theo's one-acre shed remained frost-free over the cold winter of 1901–1902.

The winter of 1901 was a cold one with about three frosts a week from November onwards, but Theo's shed remained frost-free and the young trees were looking promising and even had a little fruit on them, although they had only been planted a year. His steam pump in the boiler house supplied water at 70 degrees and 300 gallons per minute to 30 hydrants in the shed, whose spray covered the ground, keeping the inside as warm as his heated greenhouse. Close to the house, Theo erected a warning system consisting of an electric thermometer on a post near the house set to ring an alarm in his bedroom when the temperature reached 34 degrees. This gave him time to fire up and start the steam pump. After about an hour and a half, he had adequate steam pressure to produce a dense fog of warm spray, increasing the inside shed temperature to around 48 to 50 degrees, even when it was as low as 18 degrees outside. A cot in the attached boiler house allowed Theo to catch some sleep when all-night freezes threatened—see Figure 22.2.

Theo's shed again came into its own in the winter of 1904–1905, as he described in the May 1905 issue of *Country Life in America*. He declared in the introduction "those that see the accompanying pictures will conclude that all my hopes are ruined," but, of course, that wasn't the case. The article described how a great blizzard attacked Central Florida in January 1905, damaging nearly all the oranges remaining in the groves of the region and causing the fall of half the foliage. The lowest temperatures recorded by Theo for the four nights, taken five feet off the ground near his shed under a large live oak, were 20, 24, 19, and 26 degrees. The wind blew a violent gale from the northwest all the first night, tearing some of the cloth covering away from the roof and driving icy blasts down into the shed. Theo took a picture of a part of the inside of the shed the next morning and described the scene:

> In some places, where cross-drafts met, every twig became encased in ice and every leaf had its icicle, many of them a foot or more in length, making a fairy spectacle next morning of vivid green foliage and golden fruit, sparkling in the sunlight with myriads of Arctic diamonds, to whose beauty a photograph does scant justice.

When the ice melted, Theo found no injury to bud or fruit and only a small number of leaves lost in a part of the shed that had suffered the most. Thirty-five days after the ice storm, Theo took a second picture of the same area showing new growth six inches long on the trees and an abundance of fruit on the grapefruit tree by the post. By comparison, outdoor trees were in a critical condition, their tender new growth being particularly susceptible to even the lightest frost. Theo concluded that with constant overhead irrigation and his method of protection, "Shedded trees, properly cared for, are practically safe from anything but a blizzard, which rarely comes to us oftener than once in a decade."

The *Country Life in America* article contained the first known description of the process of continuous water irrigation on citrus to create a protective ice cocoon, a technique still used today in the citrus and other cold-sensitive plant-growing industries.

28.2: *Theo's shed on the morning of January 27, 1905 (left) gave the impression of the citrus as being ruined; thirty-five days later (right) there was abundant new growth.*

Shedding was clearly the way to go for tender crops, and he realized that to obtain the best prices in Northern cities, early crops were necessary when produce from other colder locations was not available. Cucumbers, eggplants, and cauliflowers were all tried with varying degrees of success, but a 1906 Christmas planting of cucumbers hit the jackpot. Over the following year, in the ten weeks from mid-March to the end of May, 1,474 baskets sold in New York yielding a total of $2,254, and the year after 592 crates of eggplant produced a profit of $325. Overall, his experiences with vegetable growing in his shed at Lake Charm were promising but of a smaller scale than he desired.

He now wanted to expand the production of tender vegetables and join the many people already growing the latest wonder crop that everyone was talking about, celery. This grew easily in the boggy, sandy soils around Sanford, Oviedo, and Geneva. But by now Theo was in his late 50s and the vigor of youth was a fading memory when it came to digging, tilling and cultivating large areas of land. He concluded that if he was to increase production and get into the celery business, he needed to be a lot nearer Sanford. There the land was more suitable for growing celery and the necessary labor and infrastructure for harvesting, cleaning, pre-cooling and transporting any crop by rail to Northern cities was in place. So in July 1909, he paid $10 for a five-acre plot (plot 12) at the Florida Land & Colonization Company's Celery Plantation in Sanford. This cooperative of around 50 plots of varying sizes was set up near the railroad with a pool of shared workers and facilities. Theo also established a partnership relationship on a shared profits basis with Cyrus Miner Berry, who owned the adjacent five-acre plots 13 and 14 and plot 45 at four and a half acres.

Berry had direct experience in the use of temporary cloth greenhouse structures to carry fruit and vegetables through the Florida winter in safety, having been the co-author of a US patent granted in 1902. The arrangement reached was for Theo to supply horticultural support across the four plots in the growing of celery, shedded tender crops such as lettuce, eggplants, peppers, beans and cucumbers, and flowering bulbs such as amaryllis.

Under normal circumstances, the shed structures at Sanford gave adequate protection, but when sub-zero temperatures were forecast wood or coal burning furnaces were employed to provide supplementary warm air heating. Making money from tender plants was entirely dependent on the severity of the winter, and diversification was essential to avoid heavy losses. Lettuce, in particular, was tricky. It cost around $300 an acre to grow and $500 an acre to ship to New York, yielding an average profit of around $105 an acre, so the grower's loss could be much more than his gain. At some stage, as temperatures plunged, plants would be lost. As a result, returns from 1908 were meager in contrast to the riches of the 1906 and 1907 seasons. However, 1911 was a profitable season for the Sanford partnership, and they shipped 210 baskets of lettuce in the week ending April 23 and 50 crates of peppers. Nineteen-fourteen was another bumper year with one and a half acres of lettuce providing two crops and bringing $3,320. By comparison, in 1915, they had 28 degrees in November in Sanford, the coldest for 40 years, and the lettuce loss was significant. But overall that year they made a little money because the 1915 mix included two unshedded acres of cauliflowers and two of celery.

CHAPTER 29

Caladium Growing with Henry Nehrling, 1885–1920

The many articles Theo wrote in horticultural journals and magazines ratified his reputation as a leading horticulturist, both inside and outside Florida. One such article on *Pancratium ovatum* (a type of white amaryllis), published in *The Florida Dispatch* in 1890, had prompted a letter from Dr. Henry Nehrling at the Milwaukee Public Museum, asking detailed questions about Theo's garden and methods of growing palms. Theo's reply began an almost forty year cordial relationship between the two of them, involving sharing plants, pollen for hybridizing, and horticultural experiences. This was consolidated in 1903 by Nehrling's move to Gotha, near Orlando, in Central Florida.

Between 1890 and 1901, Nehrling made numerous winter visits to Gotha, where he had purchased 65 acres of land in 1884, and on some of these trips he called in at Lake Charm. He was impressed by Theo's plantings and wanted to know all about which plants did well there, writing, "While visiting you I was struck with the masses of beautiful climbers on the veranda of your house and your fine palms." He bought several palms from Theo, paying in part with a copy of his recent book *Our Native Birds of Song and Beauty*, in which he paid homage to Theo's garden:

> When visiting the beautiful place of the well-known entomologist and botanist, Mr. Theodore L. Mead, of Lake Charm, Fla., large numbers of grand palms, tropical shrubs and vines attracted my attention. The veranda was transformed in a mass of allamanda, star jasmine, *Solanum jasminoides*, *Bignonia Capensis*, and Mexican mountain rose. The exquisite yellow flowers of the allamanda contrast beautifully with the scarlet flower trusses of the Cape bignonia, the rosy-red of the mountain rose, the white of the star jasmine, and the bluish-white blossoms and deep red shining berries of the solanum.

29.1: A portrait of Henry Nehrling, circa 1890, while living and working as the Curator of the Milwaukee Museum. In 1903, he moved to Gotha in Central Florida and became Theo's friend and horticultural colleague.

Henry's move to *Palm Cottage* in Gotha marked the beginning of an intensive period of cooperative horticulture between the two of them, with the focus initially on bamboos, palms and growing bulbous plants—principally caladium, amaryllis, and crinum. The relationship was friendly and not competitive, with Nehrling concentrating on acquiring as many historical varieties as possible and propagating these in large-scale plantings with a business focus, and Theo mostly interested in the creative side of hybridizing new and unusual varieties. In this way, Theo's small-scale hybridization products could be scaled up if necessary to commercial quantities on Nehrling's acres.

29.2: The front of Waitabit *showing the mass of flowering climbers around the veranda and front door.*

The caladium had been hybridized extensively both in Europe and South America before this time. It was an easy plant to cross-breed, and rewarding with varying leaf textures and almost endless color combinations of pinks, reds, greens, and whites being possible. The work of two previous hybridizers stood out: Alfred Bleu of Paris, creator of the famous *Caladium candidum*, and Adolph Lietze, a highly educated German who settled at Petropolis near Rio de Janeiro in the year 1857, and became the greatest of all the caladium hybridizers. Inevitably, with all the hybrid activity over a fifty-year period, by 1900 the number of named hybrids ran into many hundreds.

29.3: *Henry Nehrling, circa 1908, in his lath house at* Palm Cottage, Gotha, *with some of his massive collection of caladiums. At the front are the white with green-veined* Caladium Candidum, *one of the oldest hybrids dating from the nineteenth century.*

The caladium was a foliage favorite of Theo's, and had been since the 1880s when he became the first American to hybridize the species, as Henry Nehrling later recorded, "The first most beautiful hybrids of the fancy-leaved caladiums in this country came from him." For Nehrling, it was a visit in 1893 to the World's Columbian Exposition at Chicago, where he saw an exhibit of fancy-leaved caladiums in the Brazilian section, hybridized by Lietze, that inspired him. Later in that year when he visited Theo in Florida, he was delighted to find them growing freely there. By 1905, Nehrling had decided to specialize in fancy-leaved caladiums and cultivate them in commercial quantities, ordering a collection of 325 named cultivars from a German grower for $75. A year later, his objective was to set out 25,000 to 35,000 of the older varieties and 2,000 of the rarer kinds, some produced by his cross-breeding program. Before long he was planting as many as 250,000 caladiums every year, growing them in beds 200 feet long and 10 feet wide in lath houses at *Palm Cottage*. With their lush growth, tropical appearance and jeweled colorings of many forms, they were a riot of color in his garden from June to November and much admired by visitors.

The Gotha location had its ups and downs, however. In 1907, he planted 65,000 caladium seedlings but lost many to a drought. By 1913, he had 800 varieties and more than 50,000 seedlings, but in 1917 a severe freeze killed half of the 75,000 plants he had ready for sale. This devastating blow led Nehrling to reconsider the Gotha location and eventually move his garden further south to Naples to avoid hard freezes.

For several years in the early 1900s, Nehrling and Mead exchanged tubers, pollen, and hybridizing experiences and ended up with dozens of named hybrids in addition to the hundreds of historical ones that already existed. The demand was brisk for these stunning foliage plants, especially any new ones with novelty colors. They started to appear in gardens via direct customer and nursery sales from the likes of the Royal Palm Nurseries (Reasoner Brothers) in Oneco, Florida, Dreers in Philadelphia, and Farquars of Boston, where they were listed in the annual catalogs by name but not hybridizer. Since that time, more and more hybrids have been added to the list of caladium varieties and many of the older types have disappeared or been renamed. Even in 1909, Henry Nehrling's son Arno was expressing the view that "it would be a great help to caladium fanciers if someone would undertake to publish a complete record of all the varieties raised by the older hybridizers, like van Houtte and Bleu."

In the naming of hybrids, both Mead and Nehrling recognized significant horticultural figures, close friends, and family members. A partial list of caladiums where Theo was the hybridizer reads, *Pliny W. Reasoner, Mrs. Theodore L. Mead*

(*Edith E. Mead*), *Charles T. Simpson, Hildegard Nehrling, Mrs. Jennie S. Perkins, Bertha S. Eisele, Blanche Wise,* and *D. M. Cook.* Henry Nehrling is credited with, among others, *Mrs. W. B. Haldeman, Zoe Munson, Mrs. Fannie S. Munson, Mrs. Arno H. Nehrling, Mrs. Henry Nehrling, Betty Nehrling, Dr. George Tyrrell, Jesse M. Thayer, Stuart H. Anderson, Marion A. McAdow,* and *Richard F. Deckert.*

29.4: Arrow and lance-shaped caladiums were a new type developed by Mead sometime before 1910.

The two shared their triumphs and sometimes one sent a leaf of a new hybrid to the other for comments. On one occasion in June 1906, Theo sent a beautiful white, green-edged and pink veined single leaf to Henry, who replied, "The caladium leaf which you had the kindness to enclose is extremely beautiful, and there is nothing like it in my collection. … This is so unique and richly colored that I would like to propose the name *Mrs. Theodore L. Mead*." Theo agreed and named it *Edith E. Mead*, a romantic echo perhaps from his time as a butterfly collector and the naming in 1878 of *Lycaena editha* as *Edith's Copper*. This hybrid was propagated into the trade by Nehrling, and as a result, there are incorrect historical references to it having been created by him. In a similar manner, it is also highly likely the caladium *T. L. Mead* was a Mead hybrid, but named and grown on by Nehrling.

Henry Nehrling and Theodore Mead were directly responsible for the growing popularity of the caladium in the United States, but Theo quickly became bored with the limited possibilities of the fancy-leafed forms. He wrote to Reasoner concerning Nehrling's new caladiums, "They are beautiful of course but it is almost impossible to surpass previous introductions and most of the new ones can be pretty well matched by older sorts, many of which have disappeared from cultivation as their successors are pushed." This led him sometime between 1900 and 1910 to create a brand-new type of caladium with a thin or pointed leaf, rather than the conventional heart-shaped one.

He started in 1898 with three caladium species from the Sanders Company in London—*C. albanense*, a small dull-red thick-leaved specimen, *C. speciosum*, green leafed with silvery white ribs, and *C. venosum*, a plain green, fleshy-leaved and lance-shaped caladium. By crossing and recrossing with the Brazilian standard varieties, he created a race of caladiums with dwarf growth habits, having a range of narrow and sometimes pointed leaves yet still possessing the high coloration and patterns of the larger fancy-leaved forms. At the extreme in shape, he produced strap-shaped varieties with variegated leaves three-quarters of an inch wide and a foot or more in length. As a group, they had tougher constitutions, being more tolerant of direct sun and less likely to flag in a drought. Theo named this new type of caladium "arrow and lance."

In the naming of many of these new caladiums, Theo seems to have had a penchant for using the names of American Indian tribes and Seminole Indian chiefs or places. Old hybrids from the early 1900s named *Coacoochee* or *Osceola*, both Seminole Indian leaders, or *Seminole*, *Thonotosassa* or *Wekiva*, are almost certainly Mead hybrids. Theo's Christmas present to Henry in 1919 was the lance-shaped caladium, *Istachatta*, whose name was derived from the Seminole words for "man snake."

However, despite this horticultural success with hybridization, the new caladiums were not a success with customers or the trade, and even today they are the poor relative to the more well-known fancy leaf varieties. Caladiums put food on the table but little more. They needed a lot of work in their growing and took two or three years to cultivate to marketing sizes. With the plethora of hybrids available most customers wanted a cheap basic mixture to create splashes of garden color, and Theo realized that he needed a better financial return from his endeavors. So after more than twenty years growing and hybridizing them, in 1920 Theo sold off the remains of his collection of 2,600 mixed fancy caladium tubers to Burpee of Philadelphia for $75 a thousand.

CHAPTER 30

Amaryllis, 1889–1930

Growing vegetables paid Theo's bills but did not lift his soul—he was hoping for more beauty and creativity in his horticultural life than sticks of celery. He remembered the 'crowning glory' of a single large amaryllis he flowered in his garden at Fishkill when only five years old, and the desire to grow these aristocrats of flowering bulbs had stayed with him. Over the years, he always had a few in his collection and experimented with hybridizing the genus as early as 1889. He loved the decorative effect of the large, colorful blooms in a chapel or at other formal occasions, so when the interest in hybridizing caladiums began to fade, he found in Henry Nehrling someone who shared a similar passion for this handsome plant.

Nehrling already had an excellent collection and several hybridizing successes, and it was not long before he had 10,000 amaryllis growing at Gotha, mainly the species *Hippeastrum rectulatum* and *Hippeastrum pardinum*. The gift of a single bulb of one of his hybrids seems to have been the catalyst to ignite Theo's interest in growing them again in quantity, and Henry then gave Theo full access to all the pollen needed for his hybridization experiments from his collection of the best European hybrids.

30.1: Theo's handyman, Clayton Newton, in a corner of Theo's amaryllis garden at Lake Charm, circa 1910.

A quarter of a mile away from his greenhouse at *Waitabit*, Theo set out an acre or so of well-drained, light sandy loam for his amaryllis cross-breeding experiments. From his early blooms, mainly of red and white color combinations, he selected those that showed a thread border of red around the edges of petals and white centers splotched with red, and set about developing the border and clearing up the white center, in a process that would occupy him for the next ten years.

At the end of that time, by selecting desirable flower characteristics such as the lightest blooms, the darkest, and those with a hairline color, Theo had created more than 700 different varieties of the original Nehrling hybrid family. Outstanding among these was the "Mead-strain" amaryllis, a reliable and hardy amaryllis with trumpet-shaped red and red/white flowers, ideally suited to the hot and humid summers of Florida and the lower Southern states. The parentage of this hybrid is lost in the multiple crossings and back-crossings that took place during its creation; all that definitively seems to be known is that Nehrling supplied the original pollen from his collection. When Henry moved from Gotha to Naples, Florida, in 1917, he took some of his amaryllis with him but so successful was the strain that he ended up ordering hundreds of bulbs from Theo in November 1919 and January 1920.

30.2: *Henry Nehrling photographed by Theodore Mead at* Palm Cottage *in Gotha sometime before 1917.*

By 1915, Theo's collection of four or five-year-old bulbs ran into the many thousands, with bulbs averaging around a pound in weight. The horticultural challenge had been met, and now he was ready to recover some of his investment, writing, "I am anxious to get back some of the wages their care has cost me all this time." Slowly, as word got round the business grew, from requests for a few seedling bulbs at ten cents for mixed and fifteen cents for pure color from individuals who visited his garden at Lake Charm, to orders for hundreds of bulbs from nurseries and public gardens, such as those of the City of Jacksonville. Eventually, in 1918, the giant Vaughan Company of Chicago offered to buy all his amaryllis production on the understanding that he would not supply any of their other Northern competitors. Their catalog description ran:

> Giant Hybrid Amaryllis: One of the finest strains of amaryllis we have ever seen, raised by an eminent grower whose entire stock and hybridization we have purchased in its entirety. Flowers are immense, borne mostly two or three to a stem, in an arresting and beautiful range of colors, blended, splashed and striped in a manner not seen in other strains. The bulbs are all large, three inches or above and are sure bloomers. Prices, Florida grown, doz. $4.50; per 100, $35.00.

30.3: Theo in his garden at Lake Charm in April 1919 when his primary income came from selling the Mead-strain amaryllis in bulb and cut-flower forms.

The contract relationship with Vaughan's for hybrid amaryllis continued to grow in size and contributed the bulk of his income. In 1920, they wanted 3,000 bulbs for the Fall to Spring 1921 season and the same number the following year, paying $125 per thousand, so something had to be done to increase capacity. Suitable land for expansion was scarce at Lake Charm, so part of the celery plantation land at Sanford was given over to bulb production where the necessary workforce to fertilize, water and generally care for them could be subcontracted.

The amaryllis garden at Lake Charm remained his showcase where he kept choice plants and reserved anything that was outstanding for breeding. By the mid-1920s, Theo had visitors most weekends eager to view his greenhouse and gardens, particularly at amaryllis blooming time in the spring, when it was a spectacular sight and said by many experts to be the finest collection outside of government gardens.

30.4: *On Sundays in Spring, Theo generally had carloads of visitors to see his flowering amaryllis garden. Photograph dated March 30, 1925.*

On a Sunday, it was usual for him to have up to ten carloads of visitors, some coming from as far away as 140 miles. Many people arrived unexpectedly, and at odd times during the week, expecting a tour just the same. Theo was generous in the time he spent with his visitors, patiently explaining the process of hybridization and the time and skill required to bring about his spectacular creations. He took the tours in his stride and outwardly never complained, even though some of his friends urged him to put up signs to the effect that these were private gardens and visitors needed an appointment. He liked his visitors to go away with something, so he often gave away a few blooms for free or allowed visitors to select any seedling flowering amaryllis they wanted for 50¢, a process that he later admitted was generous but a bit of a commercial mistake.

30.5: Left. In 1925 the Mayor of Orlando, James LeRoy Giles (on Theo's left with an unidentified man in the center) visited Theo's amaryllis garden. Right. Several years later the visitor photographed was the horticulturist H. Harold Hume, of the Glen St. Mary Nurseries Company, soon to be at the University of Florida.

From time to time, his visitors included important political figures and horticultural experts. Probably the most important visitor, although Theo didn't know it at the time, was Edwin Grover, newly appointed Professor of Books at Rollins College in Winter Park who visited on December 12, 1926. In later years, he would be instrumental in helping to save the bulk of Mead's horticultural collection and creating the Mead Botanical Garden in Winter Park. Almost certainly accompanying Edwin on these visits was Grover's brother, Frederick, a plant lover and Professor of Botany at Oberlin College, Ohio. He recognized the importance of Mead's horticultural work and on his initial visit to see Edwin in Winter Park reportedly said, "I want to meet the famous Mr. Mead."

April was the usual flowering period when the cut-flower business was at its peak, but sufficient flowers needed to be left to set seed for propagation purposes. The collection and sowing of seed took place over the summer. In 1926, he wrote, "I took 50,000 hand crost seeds to my partnership garden last summer and had them planted celery seed in style and they made wonderful growth that summer—far better than I have been able to get in flats of rich compost. At present they are dormant but nice bulbs." November was the busy period of the amaryllis season when bulbs for sale were lifted, cleaned and express-shipped to the Vaughan's Seed Store depot in Chicago.

There was some success with the marketing of field-cut amaryllis flowers and he supplied several flower booths in the area and the Seminole Hotel foyer in Winter Park, where large vases displaying his cut amaryllis blooms attracted the hotel visitors. As a result, some of them contacted him with requests for bulbs and a tour of his gardens. Not all went smoothly in the cut-flower business, however. He found that some dealers dumped them instead of offering the flowers for sale, refusing to attempt sales in competition with their own flowers, prompting Theo to later remark "Perhaps a little more perseverance would have found a market for cut spikes as they were most gorgeous."

After five or six years of bulb sales to Vaughan's, they placed an order with him for 40,000 hybrid amaryllis seeds, together with a request for details on how to plant and start the seeds. When Theo quoted a price of $4 per 100 for the hand-crossed hybrid seeds they balked at this, calling it not commercially feasible, and giving Theo an excuse for terminating their relationship rather than providing them with a surefire way of imitating his business.

30.6: Theo sits on the steps of the overgrown porch at Waitabit with a bunch of field-cut amaryllis blooms.

CHAPTER 31

Bromeliads, Crinum & Other Flowering Bulbs, 1885–1936

Bromeliads were one of the last major plant groups that he grew and hybridized. Theo had started at Eustis with the pineapple—the bromeliad genus, *ananas*—and at Lake Charm he collected the many native "air plants"—the bromeliad genus, *tillandsia*—and grew these in the oak trees around *Waitabit*. Serious hybridization began in the 1920s when he was in his seventieth year. His friend Henry Nehrling was a keen collector and grower of the genera, sourcing them from all over the world but was content to rely on his friend Mead to do the hybridizing. They collaborated extensively, and between 1900 and 1936, they were the two key figures bringing the bromeliad to prominence in North America. "You are a great hybridizer," Henry told Theo in a letter dated 1924, and in his writings described him as a more skilled hybridizer than even the famous Luther Burbank.

The first recorded Mead hybridization event took place in January 1922 when Henry sent flower spikes from *Billbergia zebrina* and *Billbergia saundersii*, the pollen of which Theo used to cross his *Billbergia nutans*. As with his orchids, to try to ensure that future generations could trace parentage without too much difficulty, Theo created a nomenclature of hybrid naming using parts of the names of the parents. This resulted in his first two crosses with *Billbergia nutans* as a mother being labeled nu-ze and nu-sau.

Always one to try something new, Theo looked to creating the unusual by cross-breeding across genera, and started with his favorite, *Billbergia nutans*, a medium-sized green strap-leafed plant with a pendulous drooping flower stem, and crossed it with a *Crypthanus*, a low-spreading rosette variety. His notebook documents him receiving *Crypthanus beuckeri* from the Brooklyn Botanical Garden in September 1925 and crossing and reverse crossing it with *Billbergia nutans* in January 1926. Seeds ripened and were planted over that summer, and twenty-one mature plants transplanted into a box in April 1928, producing flowers in August 1928. Theo photographed "Billcrypta nutans-Beuckeri" with a *C. beuckeri* in front for comparison purposes, and hand-colored the resultant print. The result was the world's first bigeneric bromeliad—a Biltanthus, with the reverse cross to produce a Cryptbergia made shortly after that.

31.1: A hand-colored photographic print entitled "Billcrypta nutans-Beuckeri" of what is considered to be the world's first bigeneric bromeliad created in 1926, and photographed by Theo in August 1928.

Throughout the second half of the 1920s, Theo had hundreds of bromeliads of all types in his collection, and gave away or exchanged many of his hybrids to the Brooklyn Botanic Gardens, to Nehrling, to Reasoner, and many other Florida growers and collectors. During this time, Mulford Foster, soon to be the prominent name in bromeliads in North America, visited Theo at Oviedo, keen to learn as much as he could about this plant group. He was a frequent visitor, living only twelve miles away in Orlando, and through Theo's generosity acquired many hybrid *Billbergias* and Cryptbergias, which he listed after Mead's death without credit in his plant catalog. The visits must have acted as a vital stimulus to Foster's future vocation as bromeliad expert and self-styled "Father of the Bromeliad."

Theo had made good money and gained recognition with growing amaryllis and extended his interest to other flowering bulbs hoping for similar financial success. In the late 1880s, he ordered twenty-six different iris species from a grower in New Jersey, commenting, "I wish to hybridize the finest sorts and get varieties that may be quite new and unique." Later that year he started to hybridize other types of lilies, particularly *Lilium pumilum*, the Coral Lily of Siberia, whose flowers he described as looking as if they had been "carved from the reddish Mediterranean coral—or as if made of sealing wax."

He also obtained a large collection of crinum from a Mr. Lancaster, an English collector living in Lucknow, India. When the parcel arrived, it contained nearly eighty species and varieties of crinums together with some alocasias and other plants. He drained the small muck pond in front of his house, sprinkled thirty barrels of lime on the muddy surface, followed by wheelbarrows of sand to create "as fine a garden spot as could be desired." Then he planted forty-three named and half as many unnamed crinum varieties around the bed of the pond and hybridizing began the next season when they flowered.

Twenty-four different crosses yielded good seeds but at times only air filled the large pods. The most successful pods were big solid masses as large as a fist, sometimes containing as many as eighty seeds. *Crinum capense* fertilized with different kinds such as *C. pedunculatum* or *C. kirkii* often produced good pods, and out of this hybridization came the hybrid between *C. kirkii*, a species from Zanzibar, and *C. capense* that Theo christened with a compound name, *C. Kircape*.

Several years later, the drainage tiles blocked up and heavy rain flooded the old pond area, resulting in the loss of his collection, except a few bulbs he had in another location. Years later he would plant the remaining small bulbs in fertile, moist soil and after seventeen or eighteen years of growth one of the hybrid crinums surprised him by flowering—a fragrant, light pink almost white bloom, with more intense pink on the reverse of the petals, which he named *Peachblow*.

In about 1910, he turned his attention to gladioli, with the objective of developing new novelty yellow varieties and creating new strains suited to the Florida climate. He bought seed from Germany, some bulbs of the best primulus varieties, a few species, and for $5 a single bulb of the yellow gladiolus *Golden Measure* for crossing. At flowering time, he crossed the best of his varieties and species back on the parent flower for the first generation, then later crossed the results back again on *Golden Measure*.

Growing hand-crossed seed with the objective of promoting novelty and unique features was the exact opposite of commercial operations aimed at creating volumes of a standard variety. For novelty, even weakling seedlings could be valuable, as Theo wrote, "If a breeder gets a seedling with some unique and desirable character, but otherwise worthless, it is merely a matter of time and patience and judicious crossing to impose that character upon a robust and satisfactory variety." But, he added, the hybridizer "must make up his mind as to what he wishes to accomplish, keep track of his pedigrees and be willing to persevere through several generations in order to attain his ideal."

Natural pollination carried out by insects and birds hindered the process of gladioli hybridization. In the evenings, hummingbirds in their hundreds visited the flowers for nectar, poking their way into every single bloom and requiring him initially to tie small paper bags around opened flowers to prevent this normal pollination. Later he developed a system of early morning visits to remove the stamens and pollen from newly opened flowers, then making a second round hand-crossing selected flowers with the pollen. In time he obtained a selection of choice seedlings, with five-inch golden yellow flowers, growing five and six feet tall, that seemed well adapted to the prevailing climatic conditions. These plants produced beautiful flowers, some flecked and dotted with orange and coppery red. They made wonderful floral arrangements as cut flowers when accompanied by fronds of leatherleaf fern, which he had growing in the shady areas of his estate.

However, he had bad luck in the commercial introduction of new gladioli cultivars, writing:

> Within a few years I accumulated quite a stock of the best types, and a dealer in Deland, Florida, made an arrangement to grow them on and introduce them. This was about the time of the Florida boom which drew men from all trades into the real estate business. The dealer suddenly left his bulb business and after they had been growing on for another year in Deland I went over to investigate the situation. I found the bulb man's foreman had let them become crowded with grass and eventually all I obtained from the lot was half a bushel of gladiolus bulbs out of the two and a half bushels I had supplied in the first place. This discouraged me in any future attempts to introduce the types I had originated.

Theo also grew and hybridized *Hemerocallis* (Daylilies). These were easy to grow in Florida, but once again he found most hybrids did not meet his high standards of novelty or improved character. After ten years of breeding, there was only one variety, *Chrome Orange*, which he considered worthy of introduction. He imported many nerines from various sources, and those that survived did well until one night they were all stolen and never seen again. A bulb of the rare Ismene-type of yellow *Hymenocallis*, known as *H. amancaes*, imported from Haage & Schmidt in Germany and one of the rarest bulbs known had a more grisly fate. One night as it hung curing in his bulb house it was attacked and devoured completely by a lubber grasshopper. Despite these setbacks, there was no difficulty at that time in importing bulbs and plants from anywhere in the world, but it would not be long before bureaucratic quarantines would close this particular door.

In an effort to repeat his amaryllis success, he took up growing the paperwhite narcissus (*Narcissus papyraceus*), the speculative bulb to cultivate at the time. It flowered in southern gardens from December through to February and the cut flowers were highly prized by Northern florists, being the undisputed flower of choice for funeral arrangements and winter bouquets. As a money crop, the small-scale growing of these bulbs began in Florida in the early 1920s, providing winter flowers and marketable bulbs for autumn planting. Theo convinced his business partner at Sanford, Mr. Berry, to take the plunge in growing them, and they converted one of their five-acre plots to the growing of four acres of paperwhites alongside an acre of Mead-strain hybrid amaryllis, planting a further 40,000 narcissus bulbs in October 1926.

It was a spectacular success, and in May 1927 they had raised a million paperwhites and an acre of fine amaryllis. At this time, when according to Theo "the quitting was good," they agreed to an offer from Mr. L. D. Drewry of Daytona Beach, Florida, of $16,000 to purchase the entire five-acre crop. As part of the deal with Berry, Theo sold the five-acre plot at Sanford to him, making his profit from the sale more than $10,000. This ended his money worries and closed the chapter on commercial amaryllis growing except for a reservation of about seven hundred choice bulbs, which he continued to use in hybridization experiments at Lake Charm.

31.2: *Theo examining the flowers of* Heliconia psittacorum *in his Lake Charm garden sometime in the 1920s.*

Part Eight
The Final Years

CHAPTER 32

The Oviedo Boys & Scoutmaster Mead, 1920–1930

Despite horticultural fame, there was a significant unfulfilled part of Theo's life: the one child he had had died in early childhood and his desire to have a boy of his own had come to naught with the frustrating experiences of Harold's adoption. Unsatisfied parental instincts had led him on one occasion to write to his parents that parenthood for him would be the only thing that would make life really worth living. To fill this void, he acquired many pen pals over the years, typically children of friends and family, sending them little gifts particularly at Christmastime and having them reply to him as Uncle Teddy.

Theo had known Edith's younger sister Anne as long as Edith herself. He was very close to her and her children, Eleanor Dudley Smith, and Catherine Tappan Smith, who had been a playmate of Dorothy. Both nieces had four children, and as the older members of the Edwards family passed away, this new generation provided the contacts for Theo on his visits to Coalburg, as Uncle Teddy.

As well as pen pals and family children, Theo had great fun entertaining the

Oviedo village boys at *Waitabit*. He thought their shouting and laughing was the nicest noise in the world, and he was much respected by the youngsters as a highly knowledgeable and fun-to-be-around father figure. Nothing gave him more pleasure than feeding their curiosity with scientific and engineering toys and games and watching their faces as he demonstrated chemistry experiments involving smoke, smells, and explosions. He would gently educate them in botanical matters with hand microscopes and provide them with simple horticultural tasks. When it was time for them to leave, Edith would feed them homemade lemonade and gingerbread or other sweet treats that she'd baked, or gave them oranges to eat from the trees. He wrote to one of his pen pals:

> I'm just an old "grown up" who likes boy-comradeship better than most anything in the world but who never had a boy of his own. But a whole bunch of neighbor boys come and see me most every Sunday afternoon—I have mechanical and electric toys for them to play with and read them Jungle book stories and so on and I think they like me a bit as well as the toys.

32.1: *Theo's Oviedo boys loved visiting* Waitabit *on Sunday afternoons. Left: Charlie and Miles Weston, March 30, 1912. Right: An unidentified boy with Joe Leinhart (white shirt), date possibly 1907.*

With the essential elements of a chemistry set at hand, a favorite pyrotechnic was lighting a fire by pouring a few drops of water onto fine wood shavings mixed up with a speck of potassium. On one occasion, he had a bubble party on the lawn and constructed a hydrogen generator for gas to blow the balloons up with, but the boys loved to hear them go pop when touched off with a candle, so not many of them escaped up into the air.

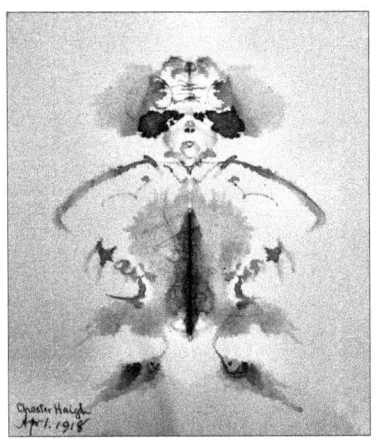 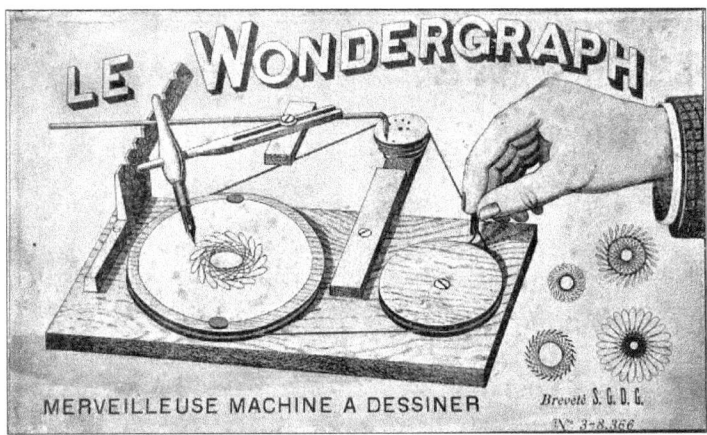

32.2: Theo was fond of amusing his Oviedo boys with toys and games, generally of an educational nature. Chester Haigh, one of his scouts, produced the Gobolinks ink pattern in 1918. Drawing geometric patterns with the Wondergraph was another popular pastime.

Constructing toys from wood and paper and seeing the look of wonder when the toy was operated also gave him pleasure, as did the production with ink and paper of *Gobolinks*, "Shadow Pictures for Young and Old," constructing geometric patterns with the *Wondergraph*, spinning his kaleidoscope top, and bending glass tubes over an alcohol lamp.

About 1920, an opportunity arose to form the first Oviedo scout troop, and Theo was the obvious choice as scoutmaster. The recreation room of the library or *Waitabit* cottage provided the meeting venues and the scouts in attendance in early January 1921 were: William R. William, Sidney Swope, Olin Wright, Allan Thompson, Malcolm Jones, Fred Ellis, J.B. Jones, Chester Haigh, Harold Varn and Ernest Kelsey.

For more than ten years, Theo would continue his efforts with the scout movement, entertaining the boys at *Waitabit* on Sundays, attending camps with his troop, and wearing his Scoutmaster hat on every possible occasion. The Scouts became part of his extended family. The troop swelled to around thirty in number and every Christmas he gave them small gifts, cards, and the obligatory Scout diary.

32.3: *With his white beard, and love of children, Theo gladly embraced the role of Santa at Christmastime.*

Among the children of the town, there was gentle controversy as to whether Theodore Mead looked more like Moses or Santa Claus, but for obvious reasons this disagreement quickly ended at Christmastime. Theo relished the giving side of Christmas and with advancing years would play the role of Santa and increasingly look the part as his beard and hair whitened. As Claire Evans, who as a young girl growing up in Oviedo knew Theo, wrote, "I recall him being such a happy man that my memory of him is of a little Santa Claus. He was not large, but he was rotund and had a happy look about him. I never saw him when he wasn't smiling and his blue eyes twinkling. He had a little Vandyke beard and Santa Claus just seemed to fit him."

32.4: *Scoutmaster Teddy Mead at Coalburg, West Virginia, in September 1922 with nephews Ogden Edwards Willis (6) and John Augustine Willis Jr. (13).*

During the summer of 1922, he took the troop to a Scout camp at Silver Lake. In a letter to Dr. Lewis Knudson at Cornell University, he said:

> I am getting ready to go off on a fortnight's camp with some of my Boy Scouts. I am in a state of frozen horror as to what will happen to my work

while I am gone, and at my age (70) I don't particularly appreciate the privations of camp life. But "Youth will be served," and anything that my Scouts want, they have got to have, if I can give it to them.

It was at this camp that Theo first met a fourteen-year-old Scout from Orlando, John Connery, called Jack by his friends, who later in the early 1930s became his young disciple, spending many weekends and sometimes longer periods at Oviedo assisting Theo with his horticultural work. In addition to his scouts, Theo's Coalburg nephews and nieces were always pleased to see their Uncle Teddy and he generally brought gifts of oranges and small toys to play with. At Coalburg in September 1922, he made honorary scouts out of the two Willis boys, John Augustine Jr., and Ogden Edwards.

For the Meads, transportation locally in the early 1920s was still via horse and buggy, with longer journeys to Sanford, Winter Park or Orlando on the train. They needed a more convenient form of transportation now that Edith was regularly going to the library and Woman's Club, and Theo was making almost daily trips to his partnership activities at Sanford, carrying gardening supplies and equipment. So in early 1923, Theo bought a second-hand Model T Ford touring car with canvas top and open sides from his friend Mr. Berry. With this purchase, delivered on his birthday, he joined the elite group of Central Florida motorists at the age of 71 with his flivver.

Over the next month or so, his handyman Clayton Newton taught him how to drive. Theo, with his engineering background, spent time tinkering with his new toy, cleaning the spark plugs, making a gasoline gauge stick, and generally performing other mechanical adjustments. Despite this engineering bent, the diaries through the 1920s show that although he understood the mechanics of the car pretty well, this did not necessarily translate into proficiency in the art of driving. In March while practicing, he bent the starter crank trying to turn and recorded, "Must practice in a safe area!" With its open sides and canvas top the car could easily be swamped by heavy rain, making starting virtually impossible, so Theo had a new garage built with a cement floor. But the garage was another obstacle. On a number of occasions he bent the fenders or damaged the canvas top maneuvering in or out of it, and there is an anecdotal story of him repeatedly crashing through the back wall of the garage, forcing him to erect a dirt embankment to stop the car.

32.5: *With Edith at the reins, Theo stands by their Murray buggy pulled by Dick the horse, outside* Waitabit *in February 1922. A year later he would buy his first car.*

In 1924 on returning from Sanford, he sideswiped another car on his way back home, breaking a wheel and axle for which towing was $10 and repairs $18.35. With the rutted and sandy roads, getting stuck and bogged down was a frequent occurrence, and getting the car started again was not always easy. "Got as tired cranking as if I'd walked ten miles," he recorded in a 1923 diary entry. Once when driving with Edith near Winter Park, he turned the car over on a particularly uneven patch of sandy road. Neither were hurt, but legend has it that Edith with her proper New England training, crawled out and dusted off her skirts, then looked at her husband and said, "Really, Teddy, we must keep a whiskbroom in this car."

Despite these little problems, overall the car made a big difference in the Mead's mobility. Trips became much more convenient to his partnership garden in Sanford, to Orlando for visits to the dentist and shopping, to Winter Park for church communion services, and to the library and back. His Oviedo boys benefitted too. They continued to make *Waitabit* their second home at weekends whether they were scouts or not, although most were. With the acquisition of the flivver, he took the boys on trips to local spring-fed lakes in Seminole County—Crystal Lake, Silver Lake and Palm Springs—to enjoy swimming, an impromptu ball game, and a picnic, generally a wiener roast.

32.6: Theo with two of his scouts on the newly thatched roof of the Sweetwater Park pavilion.

He never turned the boys away, making them fishing poles from the long, straight golden bamboo that grew on his property and generally entertaining them. But he had them working on community projects when he could, and one such opportunity arose at Sweetwater Park. Edith's position as a founding member of the Oviedo Woman's Club had been instrumental in 1922 in generously giving five acres of the Mead property along Sweetwater Creek, to create the first recreational park in Oviedo. In 1924, volunteers built a pavilion in the park to provide shade, and Theo and two of his scouts thatched the roof of the wooden structure with palmetto leaves.

32.7: *Theo and two of his Oviedo scouts, John Lawton (center) and Jack Varn, pose with an orchid in the gardens of* Waitabit, *March 2, 1924.*

Theo took almost as many photographs of his boys as he did his orchids, hand coloring many of the resultant prints and making up albums of his subjects. When he wanted to be in the shot, Edith would trip the shutter of his five-by-seven-inch plate camera. He loved to experiment and for amusement in March 1924 took two negatives of one of his boys, sitting at opposite ends of a garden

254   Orchids and Butterflies

bench, then made a mask so he could expose the two different images on the same piece of printing paper.

32.8: Theo loved experimenting with aspects of trick photography, as in this example, taken March 23, 1924.

The scout troop was precious to Theo and he wrote, "I take great pride in being the local Scoutmaster, and some of these 'good scouts' could hardly be more dear to me if they were truly my own children. They repay a thousand fold all that I am able to do for them." The annual Scout trip took place in July 1928 to Camp WeWa, an eighty-acre camp between Apopka and Plymouth, where he renewed his acquaintance with Jack Connery, now twenty and an Eagle Scout acting as a scout camp assistant.

In 1930 after around ten years in the role of Scoutmaster, at age 78, Theo stood down, stating, "I've resigned as Scoutmaster but am still merit badge commissioner and hang around with the crowd just the same. Mr. Laney the school principal is just the cat's whiskers as a Scoutmaster—thinks up jolly stunts all the time." Hanging out at the Mead place among the plants and flowers continued to be a favorite pastime for the boys, and one of them, Joe Leinhart, later became a commercial grower largely as a result of Theo's early horticultural influences.

CHAPTER 33

Modern Conveniences & Medfly Crisis, 1925–1930

In late 1925, electricity came to Lake Charm and Theo had wiring brought in from the Lake Jesup highway. A telephone was installed and he ordered an electric range for Christmas. With the car, and electricity in the home, life was getting a little easier. Eating out was becoming a regular event as well, and after church on Sundays, the *Whistling Kettle* in Winter Park was a favorite venue for dinner.

Over Christmas, an invitation arrived for him to attend the 1926 Alpha Delta Phi initiation at Cornell as a special guest, to give the "charge" to the initiates and act as toastmaster for the evening banquet. The accompanying letter added that transportation would be paid both ways. The event took place on February 20 and afterward the boys presented him with a gold badge recognizing him as an honorary member of the class of '26. "I haven't any words to express my love and devotion to these splendid brothers and our blessed fraternity," wrote Theo.

The Meads were making reasonable money from the amaryllis business, and in January 1927 took delivery of their second car, a new Fordor Sedan, paying $643 for the car and another $44 in extras—$18 for bumpers, $12 for a speedometer and $14 for a spare tire. The four-door car was a considerable improvement; it had an electric starter so there would be no more hand-cranking, and a closed cab so it didn't fill up with water like the flivver when left out in Florida downpours. With two cars, more of his scouts could be taken to ball games or scout camp. There was money for travel too, and over the summer of 1927 Theo repeated the visit he'd made the previous year to Ithaca, this time attending the fiftieth anniversary reunion of the class of '77.

Edith's debilitating headaches were becoming more common and sometimes prolonged and severe enough to cause her to take to her bed for days. She was frequently tired and worn-out. At times, she slumped into a melancholy existence sustained only by her Christian faith, waiting for the door that closed behind Dorothy all those years ago to open for her and for them to be joyfully reunited.

On Wednesday, October 19, 1927, she woke up and was half-dressed, when she suddenly complained of a terrible pain in her head and lay down and was unable to lift herself, even to get the sheet back over the bed. Theo telephoned Dr. Samuel Puleston of Sanford, who came, diagnosed a cerebral hemorrhage, and organized for a nurse and ice caps. The following day she was much worse and unable to speak. The doctor came twice but on the last visit said there was no hope of recovery. Edith passed away at 7:30 in the evening of October 20, after which Theo telegraphed the news to Webster at Coalburg.

The next day a neighbor took him to Orlando where he made funeral arrangements with the Carey Hand Funeral Home and bought burial lot number 119 in section B of Orlando's Greenwood Cemetery. Mrs. Annie Lee Carter (née Lee), mother of Walter Carter one of his scout troop, took him home for the evening and insisted he spend the weekend with them. The funeral service, conducted by Dean Gilman, took place on October 22 at 4 p.m. at the Orlando Episcopal cathedral, decorated with many floral tributes with Edith's favorite flower, the rose, predominating. By request, Edith's friend Mrs. Whitman and a small choir sang her favorite hymn, "My Faith Looks Up to Thee."

Making his final goodbye was hard, as Theo recounted in a family letter to Edith's sister Anne, "I was almost afraid to take a last look at my dearest dear, but her face was sweet and placid and she looked more like her mother than I have ever seen her before." The next day, Theo wrote a few death notices for the newspapers and letters to members of the Edwards family where he confessed, "Everybody is as kind and sympathetic as possible but with Edith gone life is but bitter dust and ashes, and I don't care how soon I am permitted to go to her." The following day, with a bad cough and a temperature of 103 degrees, he was admitted to the Fernald-Laughton Hospital in Sanford, where he stayed for the next fifteen days until well enough to be discharged on November 9.

The tributes for Edith within the town of Oviedo were heartfelt and genuine. Many remembered the work she did administering medication with sincere concern for the sick, particularly among the poorer children. Others recalled the influence she had culturally on the community, setting up the town's first library within the Woman's Club, and gifting a beautiful walnut glass-fronted bookcase and books at a time when books in the average home were a rarity. She brought music to the community, not only through her organ playing in the church on Sundays, but also through the lessons she gave to a number of local children who became proficient pianists as a result. One fond memory related to the time she offered to give a singing canary to any student who memorized and could play a certain number of classical pieces. The canaries, imported from Germany and taught to sing there, were highly prized by the children who won them, and could sing along with particular pieces of piano music, not just tweeting their own little songs but harmonizing with what was being played.

Her community spirit and love of children lived on in Sweetwater Park, and when the town built a 50-foot long swimming pool there, which opened in 1930, it gave the children of Oviedo a rare opportunity for fun and play at a time when there was little else to do. The admittance charge was about 25¢ and there would be 25 or 30 children splashing around in the pool on a busy day, which would have delighted Edith if she had been there to see it.

Back home despite being a credible cook and bottle washer, Theo found life difficult without Edith, particularly as Christmas approached. He wrote to his niece Catherine at Coalburg, "Here in this isolated place I don't know what to do about Christmas—Edith knew exactly what everybody liked and wanted and took the responsibility (except for consultation) off my shoulders." Catherine immediately asked him to Coalburg for Christmas, but Theo had to reluctantly decline, with grateful thanks, citing his $450 doctor's bill, tax bill and repairs to the car as reasons. He wrote back "It would be best of all to be with you dear people. Johnny's talk and Ogden's violin—and his engine—and the radio and all would make real Christmas cheer while the girls would add beauty and enchantment to the scenery!"

Just before his discharge from the hospital, Theo had a visit from Clayton, who cheerfully told him that he'd been running the flivver for milk delivery and other errands unknowingly without a drop of oil, which had resulted in a stuck piston and a broken connecting rod. He followed up by stating that the car was in the garage waiting to be fixed, and offered the excuse that as long as he could see two or three drops of oil at the bottom of the glass oil gauge, he didn't think he needed to get any oil. "He can do as much damage in two weeks as I could in a year of carelessness," wrote Theo.

Now that he owned two cars, Theo had no more luck driving the sedan than the flivver, although not all incidents on the road were his fault. Driving an old Cornell friend back to Winter Park, he recounted, "I sounded my horn to pass an orange truck and speeded up to pass when the cuss made a left hand turn without warning right in front of my car. I jammed on the brakes and got off with a marred fender and bent back headlight and other small damage. The truck was so high that it missed the bumper altogether."

The road conditions after heavy rain could also be treacherous with mud, rutted roads and standing water providing the obstacles. "Saturday had to take wash to washerwoman and there was a big puddle in the road so I took a chance and got stalled right in the middle of it. Scouts came along and one got a rope and big brother's car and pulled me out," read one letter entry. The narrow drive to the road from *Waitabit* was also a challenge, as he found out one Sunday morning, "Started for church but car bogged down in the driveway. Took me three hours to get out and another hour to clean the mud off the wheels." Despite these setbacks, Theo enjoyed the independence and convenience of car travel but he always took a highway map along with him. Over Thanksgiving 1928, he visited the Reasoners in Oneco, Manatee County, making a side-trip to Naples to see Henry Nehrling, and spent New Year with the Inness's at Tarpon Springs.

33.1: Theo on one of his visits, holding a flower of Allamanda cathartica *(golden trumpet vine). Photograph circa the early 1930s.*

At the start of 1928, Mr. Drewry had presented Theo with a check for $325 and a promissory note for $10,000 as settlement for the narcissus and amaryllis bulb collection at Sanford. He explained that he needed a little time to raise the cash from the property he owned, so the note included interest at 8% per annum for the period of deferment. Theo was unperturbed, recognizing Drewry as a man of honor and sound financial standing, and happy for the time being to receive the interest in the form of a $200 check every three months. Unfortunately in February 1929, Drewry died from a heart attack leaving considerable debts that became the subject of a major lawsuit. Theo was worried about whether he would receive his money and consulted a lawyer; the case dragged on, but he finally collected the settlement in May 1931.

In June 1930, Theo was informed of the death of "my own and only nephew Bill," in a two-seater plane crash near Schulte, west of Wichita, Kansas. Bill Edwards was returning from Palo Alto to New York with a companion, Leopold Schumacker, a Swiss student at Leland Stanford University, when the airplane ran out of fuel. They attempted to glide in for a landing in a wheat field but the landing gear clipped a hedge top, and the plane nose-dived into the field. Further Edwards family bereavement took place the following September with the passing of Edith's sister Anne, after a series of strokes. Theo made a visit to Coalburg over the summer of 1930 and saw her a few weeks before she died, describing her as "very ill."

Cold weather and freezing temperatures were not the only dangers for citrus growers. Most had recovered from the freezes of 1894, 1895 and 1899, and the early part of the citrus harvest of 1928–1929 had been a good one. But with almost three-quarters of the crop picked and shipped, another calamity took place. In early April 1929, an inspector found maggots in an Orlando-sourced grapefruit, and shortly after there was a report of excessive fruit drop and the presence of larvae at the 40-acre Hamlin grove located at Marks and Mills Streets, Orlando. The larvae were subsequently identified on April 10 as the Mediterranean fruit fly (*Ceratitis capitata*), and emergency inspections then found further infestations in Lake, Orange, and Seminole counties, with a dense concentration around the city of Orlando.

On April 15, the Florida State Plant Board formally announced the insect's arrival and an emergency state appropriation of $50,000 to fight it. They defined the initial quarantine rules which consisted of identifying infested zones (zones 1)—areas one mile around the boundaries of each infested property, and protective zones (zones 2)—areas nine miles around the borders of each infested zone. An emergency transfer of $4.25 million in federal funds quickly followed and all efforts galvanized under the leadership of entomologist and Plant Commissioner of the Board, Dr. Wilmon Newell.

Newell's strategy to defeat the medfly was to destroy produce and host plants to eliminate possible breeding spots, and liberally spray host trees with lead arsenate and bait them with an arsenic and molasses mixture. To carry out this program, a large field force of initially 2,900 men which grew to over 5,000 was needed, involved in gathering host fruits and vegetables, uprooting any suspected host plantings and spraying the baited insecticide.

With his entomology background, and believing this level of activity to be the equivalent of taking a sledgehammer to crack a nut, Theo attempted to inject some science and common sense at the start of the process. In a letter to the Editor of *The Sanford Herald* in May 1929, he pointed out that the Mediterranean fruit fly was an old and familiar resident of thirty years' standing in the Sanford area, and had caused negligible damage to the citrus industry on account of the prevailing climate conditions which seriously limited its multiplication. He concluded:

> If our Quarantine Commissioner had been anxious to find the facts of our previous infestation, he would have found $250,000 an ample sum to destroy the pest instead of the five million demanded, together with a million dollars of wholly unnecessary proposed destruction; when one mile of protective zone is all necessary for every practical purpose to destroy the fly, whose menace in this part of Florida has been shown to be practically negligible though possibly more serious in the extreme lower part of the State.

 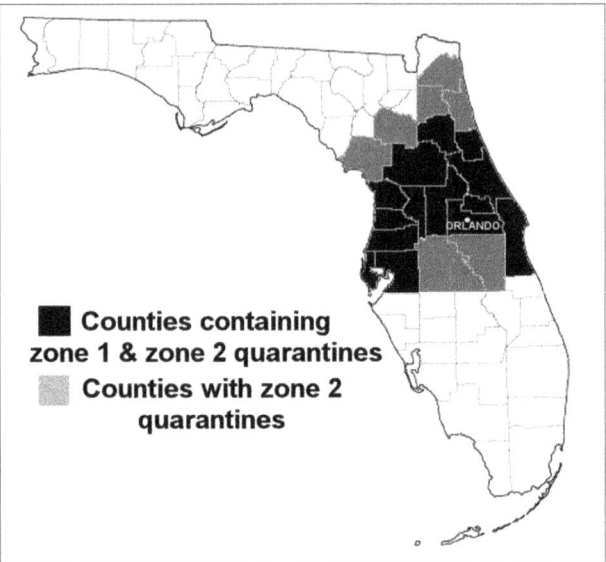

33.2: Left: The medfly crisis as seen from California in a cartoon by Edmund Gale for the Los Angeles Times. Right: The final quarantine map of Central Florida for 1930.

But the strategy of the administration remained unchanged. The unskilled, untrained and mostly uncaring laborers hacked, destroyed and sprayed their way across an area that by early 1930 extended to twenty Central Florida counties, covering 120,000 acres or about 188 square miles. People transporting infected produce was one reason why it was such a large area, as revealed in an analysis showing the infection was progressing along the main roads of travel. Consequently, the National Guard set up roadblocks along highways leading from protective and infested areas and examined all vehicles for produce, confiscating all fruit even if they found only a few tomatoes as part of a picnic.

Theo was at the heart of this devastation, living about 14 miles northeast of the initial site of infestation in Orlando. There were many infested zones within the Oviedo citrus growing area and the whole of Seminole County had infested or protective zone status. The ban on the transportation of all fruit or vegetables, except celery, and the destruction of all fruit affected the livelihood of hundreds of growers. The loss of citrus income and resultant panic among investors led to a banking collapse. "Florida banks have been exploding in the last couple of weeks like a pack of firecrackers," wrote Theo in one of his letters.

Theo's response to the mass destruction of plant stock was to make up two large signs forbidding destruction of trees and shrubs on his property and erect them at his gateways. He continued to bombard the newspapers with letters and appealed directly to the Agricultural Board, pointing out that the eradication process was having the same outcome as letting the fly multiply—the destruction of the citrus industry. However, he was astute enough to realize that "At the same time if the eradication money stops altogether, forty-seven states will quarantine everything in Florida and destroy all our industries."

The official declaration of medfly eradication took place in July 1930, at a final cost of $7.57M. Many believed that to get there Newell's scorched earth eradication methodology had been high-handed and conducted without regard to the environment or to the economic effects on growers. For Theo, who had spent his life trying to explain things scientifically and as a result acted in logical ways to events, the hysteria that accompanied the medfly outbreak, its financial cost, and the brutal excesses of the eradication workers destroying crops at times that had no connection with the fly, was one of his most frustrating experiences.

CHAPTER 34

Awards, Recognition & Culmination, 1927–1936

In 1927, the Florida Federation of Garden Clubs had adopted a resolution recognizing Theo's influence on the floriculture of Florida. They offered thanks and appreciation for a man that in the evening of his life "Has devoted more than forty years to the propagation and production of rare and beautiful flowers, ornamental plants and the like; proving, also, the adaptability of Florida soil and climate to the successful growing of flowers and ornamentals. This work eminently has been in line of making of Florida, in reality and in truth, a 'Land of Flowers'."

Theodore Mead, Henry Nehrling and David Fairchild were the three outstanding botanists of Florida in their day. In 1928, these three joined J. K. Small and H. Harold Hume as members of the Arboretum Committee of the Florida Federation of Garden Clubs, where in March 1929 they posed for a photograph.

In his hometown of Oviedo, there was recognition too. On Arbor Day in January 1931, the Woman's Club and the Parent-Teacher Association planted a tree in his honor at the Oviedo school. Mrs. Ednor Curlette of Geneva, president of the Garden Club of Sanford, spoke of his remarkable contributions to the plant world. The school composed an original song for him and sung it, and the tree was planted by the Boy Scouts, with Mr. H. J. Laney the Scoutmaster and all the scouts putting in a shovelful and Theo putting in the last. "Everybody very lovely to me," remarked Theo.

34.1: *Members of the 1928 Arboretum Committee, photographed in Miami at the Florida Federation of Garden Club's Convention, March 1929. (L-R): J. K. Small, H. H. Hume, H. Nehrling, D. Fairchild and T. L. Mead.*

As well as receiving tributes himself, in 1931 Theo was asked to reflect on the passing of Henry Nehrling, who died in November 1929. The Florida Quest Society held a memorial tribute at Rollins College on November 30, where Theo delivered a eulogy entitled "Dr. Nehrling, as I Knew Him." He drew a parallel between Henry's life and his own—both had been dedicated to horticulture with all its ups and downs yet neither had benefitted from any lasting commercial success. He talked about Henry's generosity and painted a pathetic figure of him in his final years, living on bacon and grits and "reaping few of the benefits that he had so plentifully bestowed on others, let down by selfish people who failed in their friendship towards him."

Theo's last initiation ceremony at the fraternity at Cornell had been in 1927, but as a well-loved older member of the Chapter, he regularly received invitations to this annual event. The invitation that arrived in 1932, however, was a special one and one he could not refuse. It asked for his attendance at Hamilton College for the centenary celebrations of the founding of Alpha Delta Phi by Samuel Eells in 1832. Theo departed from Sanford on August 31 by train and arrived at Utica, New York, on Friday, September 2. The Saturday banquet, with President Bruce Barton in the chair, was an evening of many speeches, songs, and fond recollections. The chapter sessions took place on Sunday where, as principal guest of honor, the 80-year-old Theo headed the "walkround" made up of around 175 attendees. After this, he visited Clifton Springs and Ithaca before meeting up with the Willis and deGruyter family members at Coalburg.

Despite his advancing years, Theo's hybridizing quest for a white amaryllis continued, and in November 1929 he talked about his journey to this objective at the Orlando Amaryllis Show. He nudged closer to his dream in March 1930 when he entered a white amaryllis with the thinnest of thin red edges at the Orlando Flower Show, which received the name, *T. L. Mead*. It took until 1935 before he eliminated the red edge and produced America's first all-white hybrid amaryllis. The flower caused a sensation among growers and he quickly sold his entire production for $5.00 per bulb to John Scheepers of New York, a flower bulb importing company with strong Dutch connections. After that, white amaryllis of Dutch extraction became commonplace and a favorite flower for Easter blooming. In 1935, in recognition of his achievements, the American Amaryllis Society elected Theo as an honorary Fellow of the Society and published a short autobiography in the Yearbook.

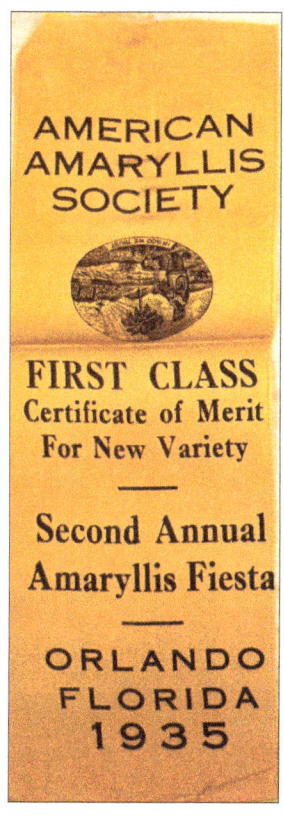 

34.2: America's first all-white hybrid amaryllis won Theo an award from the American Amaryllis Society show in 1935 where he exhibited a number of other prize blooms.

As well as amaryllis, Theo's love affair with orchids remained steady, even though he had once written, "I occupy myself also in raising hybrid tropical orchids from seed. It is the most difficult branch of horticulture and so far has yielded only enjoyment to me and no money return." Orchids then were so unusual that many people had never even seen one before, and visitors touring his greenhouse were left gazing in awe at his exotic collection. Most were happy to buy a fifty cent amaryllis bulb but not confident enough to buy an orchid, which probably suited Theo since he almost couldn't bear to part with his creations, despite the need to pay the bills. On one occasion, an orchid expert in the visiting party, stunned by one of his hybrids, offered him $500 for it, a fortune at the time. But beauty trumped cash and Theo just couldn't accept it, "Ah, thank you," he said, "but it is too lovely. I have to keep it."

The orchid appears to have symbolically represented his life, as he once wrote of epiphytic orchids, demonstrating his mastery of prose:

> They have left the vulgar competition of more earthly things and established themselves among the win'ged fairies of the air, visited only by butterflies and bees and such ethereal creatures. But you must remember that no real solidity of character can be acquired by a man or plant, without effort and struggle and the wresting of the good from surrounding evil.

34.3: Theo with visitors in his greenhouse, circa the 1930s, and with one of his Cattleya orchids outside Waitabit *in January 1932.*

Although in his 80s, he continued to buy orchids, hybridize new ones and exchange experiences with growers, but the promise of riches from the sales of orchids did not materialize. Orchids were still a rich person's hobby, and Florida was a long way away from the auction rooms of New York and London where high prices could be obtained for unusual orchid hybrids, if one collector was bidding against another.

Bromeliad hybridization gave Theo much satisfaction and he exchanged plants with California growers. By the early 1930s, his bromeliad hybrids had arrived there and entered the garden trade, and in the 1940s, Mead's Cryptbergias were widely distributed in Florida, California and other parts of the USA. Sometime in the early 1930s, Harry Smith, a nurseryman in nearby Winter Garden, Florida, bought the bulk of Mead's bromeliad collection. In the summer of 1938, Julian Nally in turn purchased many of Mead's bromeliads from Smith, mainly *Nidularium* hybrids with a sprinkling of *Billbergia* species and hybrids. Nally had earlier bought the old Nehrling *Palm Cottage* property in Gotha and used this starter collection to expand his bromeliad activities, growing them by the acre. In this way, traces of Theo's hybridization efforts survived at Gotha and were widely acknowledged and recognized by Nally.

Despite being the first American to hybridize bromeliads, Theo's old-fashioned modesty prevented him from crowing about it. When it came time for him to write his autobiographical article for the American Amaryllis Society Yearbook in 1934, he collapsed his horticultural efforts with bromeliads into a few lines, "The bromeliads interested me greatly and over the years I introduced many representatives of several genera, viz., *Aechmea, Ananas, Billbergia, Cryptanthus, Guzmania, Hohenbergia, Nidularium,* and *Tillandsia*. Work with these gave many interesting crosses of rare beauty in leaf forms and markings and in their gorgeous flower spikes." Today, only two bromeliads, *Billbergia* 'T. L. Mead' and *xCryptbergia* 'Mead' have survived out of the many new forms he created.

Visitors continued to pour into his gardens and greenhouse on a daily basis, but Theo did not complain, "They use up lots of my time but it is pleasant to have them come and they seem to be well entertained by the tales I tell them." Dignitaries were generally treated afterward to a picnic lunch by the shores of Lake Charm.

34.4: Theo, closest to the car, and some of his distinguished garden visitors, circa the early 1930s.

Throughout 1932 and 1933, his old scouting friend Jack Connery was a frequent visitor to *Waitabit*. Scouting and ornithology had brought them together in the 1920s, and after a spell as official photographer to the initial exploratory deep-sea descents by William Beebe and Otis Barton in their spherical steel vessel, the Bathysphere, off Nonsuch Island in Bermuda, Connery returned to Orlando in 1930 and renewed their friendship.

Although he had no money, Connery approached Rollins College with a view to studying there. The resultant financial impasse was eventually broken by an agreement that Connery would donate his extensive collection of bird's eggs and nests to the Thomas R. Baker Museum of Natural History at the College in exchange for tuition, on condition that he assisted with ornithology classes and became Student Curator to the Museum. It wasn't long before Connery met Dr. Edwin Grover, and they discovered their shared interest in botany and common admiration for Theodore Mead.

In 1932, Theo was 80 and Jack Connery 24, but despite this age gap a binding friendship and master-pupil relationship developed. Jack became his willing assistant and helped repot orchids in the greenhouse, collect caladium seeds and on one scout meeting evening gave a projector presentation on birds. He began to document Theo's orchid collection, made colored lantern slides from Theo's orchid negatives, and started to take parties of Rollins students to see Theo's gardens and greenhouse. Recognizing Theo's declining mobility, he also drove him to various events, meetings and outings, sometimes with the help of his parents.

Theo was invited to dine with the Connerys in Orlando on many occasions, and on February 23, 1933, they gave him a surprise birthday party at their house, "Mr. Connery came for me in the evening and I was the guest of honor at his Orlando home—15 guests—81 candles—and a medallion photograph of myself at every place-card which I had to autograph afterwards."

34.5: One of Theo's signed medallion place-cards on the occasion of his 81st birthday dinner party at the Connery's home.

Jack met his future wife, Helen Golloway, also a Rollins student and assigned to the Baker Museum, and they were married in 1934. In subsequent years, she became personal secretary to Grover, by then Vice-President of Rollins College, and would become a prominent supporter of the establishment of a botanical garden for Theodore Mead.

Theo's general health continued to be good, but a prolonged period of rheumatic jaw pain in 1929 had him visit Dr. Whitman, his dentist in Orlando, for exploratory X-rays. The dentist found some small abscesses and "recommended extraction of all teeth from upper jaw—about a dozen of them. He took out five or six there and then and said to come for the rest when I felt able." Over the next few years, he had the rest of them out and an upper jaw plate fitted. His ears were troubling him too, and he first noticed he had difficulty in making out some of the words when listening to sermons in church. Initially, he concluded that the preacher's enunciation must have been bad since others said they couldn't hear clearly either.

In 1930, he had traded in his old Victrola for a new model combination electronic radio-phonograph that came with a pick-up, giving superior fidelity and greater volume. It proved to be a great help in compensating for his loss of hearing. In the evenings, he enjoyed playing his old favorites such as operatic music by Gluck and songs sung by Louise Homer, Ernestine Schumann-Heink, and Enrico Caruso.

By 1931, he had acquired a hearing horn, which improved things a little at church, as he wrote to Catherine, "I had my 'hearing horn' along but when the minister yelled, the words were unrecognizable to me—tho' at medium voice it was a decided help. At the church door he said I had looked mighty solemn, but I told him you most always looked that way when you had had your teeth out."

The minister liked to print up short biographic sketches in praise of regular and venerable churchgoers, summarizing Theo this way:

> The man who outranks all Oviedans in the uniqueness of his personality and a wide reputation for abilities of the highest order is Theodore L. Mead. In the community his name is a synonym for every public and private virtue. No man ever knocks at the door of his heart for sympathy and finds his compassion congealed. He not only believes in freedom of speech but is fearless in expressing his convictions and conclusions. Aside from his work as a specialist in scientific lines, his hobby is forming human character on models of the noblest idealism. His influence on the Oviedo's young men will linger long after he has gone to the Great Reward. He is a model of public spirit and an exemplar of religious faith and practice.

Theo sent a copy of the sketch to Catherine at Coalburg with the comment, "When this is boiled down it means that I am an amiable cuss and love most all the boys because I can't help it."

His hearing got progressively worse and began to affect his social and scout activities, and for the remaining years he needed the assistance of a Sonotone hearing aid. His Oviedo boys were still regular visitors, but he admitted, "The boys come around quite frequently and are a great comfort tho' I cannot entertain them very well being so deaf."

When he received his 1935 invitation to attend the Alpha Delta Phi initiation ceremony at Cornell, he reluctantly had to turn it down, "It isn't much fun to go anywhere if a fellow can't hear anything," he wrote. He replied to the invitation saying how much he regretted missing the event, "With the infirmities of age coming upon me, I can hardly hope for the happiness of another initiation ceremony. ... I'd be glad to trade the remaining years, if I had them, for such a privilege." He continued, "As I wrote to the boys nearly fifty years later, the day of my initiation stands out in memory as one of the three wonder days of my life ... with my wedding day and the day of my confirmation (though formerly an agnostic) in the Episcopal church."

Increasingly, he was giving up driving to his loyal gardener/chauffeur, Clayton—loyalty that Theo reciprocated. Clayton was a quick learner and over the years had picked up significant horticultural knowledge from him, so much so that Theo not only loaned him his mortgage but the money to purchase a five-acre celery plot at Sanford, where Clayton made a $300 profit from the sale of celery in 1932. Theo's generosity to Clayton extended to him giving him the old flivver in 1934.

Theo wrote his will in August 1933, leaving his horticultural life's work of bulbs, orchids, and greenhouse plants to the Trustees of the Royal Palm Park on Paradise Key, Florida, then owned by the Florida Federation of Women's Clubs. The bulk of his personal effects he bequeathed to his two West Virginian nieces, Catherine and Eleanor. Exceptions to the bequest were books not desired by his nieces (novels and poetry to the Oviedo Public Library, histories and travels to the Sanford Public Library), and all his scientific pamphlets, apparatus and album of colored orchid photographs to the Trustees of Rollins College. John Augustine (Gus) Willis, Catherine's husband, was to be sole executor of the will.

He specifically requested that the two brown Japanese jars from the parlor at *Waitabit* were given to Julia Inness, that his marble and malachite clock went to John Augustine Willis, and that his carved gold watch and ruby chain was assigned to his godson, Theodore Mead Robie. He made sure Clayton Newton was looked after by stipulating the cancelation of any debts owed to Theo, whether for a mortgage or for advances.

Theo's intention to donate his greenhouse plants to the Royal Palm State Park took a deflection in December 1934 with the news that a severe freeze had killed some of the existing vegetation. As a result, he changed his mind over the fate of his plants. During the next year, he made several commitments, promising a representative collection to Jack Connery in one letter and the remainder to Mr. Clifford C. Cole of Coconut Grove, Florida, in another. Unfortunately, he never changed his will. This led to complications about ownership of the plants following his death, when the various parties produced letters and other evidence of his last wishes.

Theo had another dental appointment in late December 1933 with Dr. Whitman in Orlando, who told him that six of his remaining teeth needed to come out. Complications from these extractions resulted in a hospital stay and a period when Theo was confused, delirious, and "not himself." He convalesced for several weeks under the care of a trained nurse and was visited regularly by Mrs. Berry, the wife of his business partner.

After Theo's discharge from hospital, the Willises and the Berrys made arrangements for the nurse, Miss Bertha Dixon, to provide regular home care and household support and asked Clayton to take on more duties like driving the car, milking the cow, running errands, growing vegetables for the table and doing any heavy lifting work. The sickness had left Theo with an aching head and neck, although when he visited Dr. Puleston, his blood pressure had returned to normal. Mrs. Berry had noticed a difference in him, "Looks old now after hospital stay, never did before his illness," she wrote to Catherine. Even Theo recognized the change within himself, "I remain about as usual, possibly a little less torpid than before, but without hope of regaining my faculties as they were before the dentist muddled them up."

Nurse Dixon proved to be a neat and tidy authoritarian figure and set about organizing things, tidying up the house clutter and throwing much of it out, much to Theo's annoyance, "There is practically no single thing that I am specially interested in that has not been either taken away or destroyed but I have got back one of my five scrap books. … It's a big nuisance but has to be accepted. If a fellow takes temporary leave of his senses nobody can be considered responsible for what may happen to his pet things."

Miss Dixon was all for organizing Theo as well as his things, warning him of the dangers of staying out too long in the hot sun, forbidding him to eat meat, and even sending him to bed early when he wasn't tired. As time wore on, things deteriorated, prompting Mr. Willis to write to Mr. Berry expressing his concerns, "I am not at all satisfied that Miss Dixon is the proper person to look after Mr. Mead. She seems to be making him unnecessarily miserable." The situation resolved itself in late 1935, when Miss Dixon suffered a breakdown and Mrs. Carter found a local replacement in Libbie Wainright, much to Mr. Willis's relief. "I am much relieved that you have gotten someone to look after Mr. Mead. It looks as though Mrs. Wainright is just the proper person," he wrote to Mrs. Carter.

Theo had always tolerated indiscriminate and rampant plant growth. Now unchecked, the many plants, shrubs and climbing vines he had planted and nurtured around his property began slowly and inexorably to take back their rightful place as the only real long-term residents of the planet, swallowing up what small space remained around and over his house.

34.6: The cottage and garden at Lake Charm were becoming increasingly overgrown by the early 1930s.

Visitors to Waitabit in the early 1930s, of which there were many, would have been intrigued initially by the entrance drive leading to the house, heavily shadowed by luxuriant growth with its sidewalls defined by the metallic sheen of the stiff leaves of the ancient sago palms and the graceful stems of the Japanese arrow bamboo. Then as the house came into view and the vista opened out, spellbound by the variety of plants, shrubs and trees surrounding the property. Cactus and night-blooming cereus climbed up the tall palmettos and masses of bromeliad air plants and flowering orchids decorated the huge spreading branches of the moss-covered live oaks, giving the effect of an additional secondary aerial garden.

Closer to the house they would have seen a grove of golden bamboo, clumps of giant agaves and cycads, a large specimen of the palm *Acromia totai*, and an acacia tree from tropical Asia, *Albizzia lebbek*. Their overall impression standing in front of the house would have been of a mass of exotic vegetation that appeared to hold the old cottage and greenhouse tightly in its grasp. At this stage, they would have been quite unaware of the tree-shadowed road that wound from the back of the house for a quarter of a mile or so, following a ridge of land covered with open growth of hardwood trees and lined with silver-edged ferns.

A further surprise awaited any newspaper and magazine writers eager for a story when the main actor made his entrance against this backdrop of tropical plant splendor. One such reporter for the Sanford *Herald* wrote in 1931 of the experience:

> When he came out to greet us Tuesday afternoon, I had the curiously elated feeling that one sometimes has when the curtain rises on a lovely setting of a play and the hero comes out and measures up to the rest of it. Theodore Mead, with the white beard, one of those vibrant affairs that seems to share the sentiments of its owner, his ruddy cheeks, his clean cut features, and, best of all, his keen kind, sparkling brown eyes, will be eighty on his next birthday, and I think that he learned another secret from Peter Pan besides the right kind of house to live in. All about him one feels the aura of the joy of living.

34.7: Closer to the house, a mass of exotic vegetation over Theo's porch and front door greeted visitors to Waitabit.

Getting to this point, and finding the overgrown front door, was sometimes difficult for out-of-town visitors, as a reporter covering the "Plant Wizard of Oviedo" discovered:

> After several inquiries, which usually elicited such delightful replies as "Hit's quite a fur piece," I learned the general direction of Mead's place. A negro napping under a roadside palm awoke long enough to show me precisely the vent in the jungle that gives access to the bamboo alley, completely arched overhead, half dark and cellar cool. Follow this perhaps a quarter of a mile and you come out into a small open space flooded with sunshine. There you are in front of Mead's door. The wall of verdure before you gradually turns itself into a cottage, banked and propped with palmettos and leaning palm trees and knitted over with vines. … You see some steps leading up to a screened door, offering barely room for passage between masses of plumbago covered with pale blue blossom. The screened door opens. A smiling little man pushes aside streamers of climbing fern and

begonias, comes out to greet you. His square-cut gray beard, while not unkempt, plainly is not broken-spirited from excessive care. It bristles and weaves as the muscles of his cheeks work into a smile. Perfect, you think, that such an appropriate person should issue from a house just like this!

Living in his jungle cocoon, Theo occasionally walked out through the narrow bamboo alley to the road and over to the village store and post office in Oviedo. In the evenings, he continued to consult his extensive library of books to keep his mind sharp and read and reread the letters from his friends and penpals. But most of the time he was alone with his thoughts; in his mind, he continued to return to those happy Cornell years in the company of his cherished Alpha Delta Phi brothers, to him "The very ideal of brotherhood and fraternal love."

34.8: Theo in his final few years.

In early 1936, Julian Nally, the new owner of the old Nehrling property in Gotha, who was keen to understand the horticultural influences of both Nehrling and Mead, wrote to Theo asking whether he could visit to talk about future plantings at the Gotha garden, and recalled the experience of this pilgrimage:

> At his invitation I visited him and walked with him over the tangled ruin of his property adjoining Lake Charm near Oviedo. The medium of pad and pencil for questions is a difficult means of getting to know anyone and I left that afternoon more conscious of Mead's peculiarity of dress than anything else. He had on a stocking skating knitted cap, the jacket of a Boy Scout uniform and a pair of shorts over long, wool underwear.

Nally left in awe, feeling terribly ignorant, and thanked Theo for the visit in a letter:

> In the debased idiom of our day, let me say that I had a "swell" time this afternoon visiting with you. I can't help but envy you—heaven knows, envy has been put to worse uses than that toward which I am employing it! I envy your knowledge—your possession of so many rare and lovely plants, but most of all, your disposition and attitude toward life.

Nally never saw him again, although Theo continued to take a lively interest in horticultural work, recording his last bromeliad hybridization success on April 14, 1936, with the first flowering of a bigeneric cross between *Nidularium fulgens* and *Aechmea hystrix* (now *ornata*).

Only ten days later, on April 24, 1936, Theo suffered a massive stroke at his home. Within 36 hours, he was taken to the Fernald-Laughton Hospital in Sanford, but by then damage to the brain was already irreversible and multiple debilitating paralyses had set in. Alone and confined to his hospital bed, he died peacefully in his sleep on May 4.

Funeral services took place at St. Luke's Cathedral at Orlando, Florida, at

4:30 p.m. on Wednesday, May 7, 1936, officiated by Dean Melville Johnson. Six ex-scouts bore his casket, J. H. Connery, Emmett Kelsey, Arthur Partin, J. B. Jones, Jr., Ewan Jones, and Allen Thompson; scouts from the early 1920s but now themselves all in their mid-twenties. As befitting a renowned horticulturist, there was a mass of floral tributes—eighteen in total—from friends, relatives, "The Oviedo boys," the town of Oviedo, and Alpha Delta Phi at Cornell. Interment took place at the family grave in Orlando's Greenwood cemetery, with the death certificate recording his age as 84 years, 2 months, and 11 days.

34.9: *The Mead family grave in Greenwood cemetery, Orlando, lies beneath a tall pine tree where eagles nest each year.*

CHAPTER 35

Aftermath & Legacy, 1936–1940

"Science loses Dr. T. L. Mead—World Known Oviedo Botanist Dies" headlined the Orlando *Sentinel* on the morning of May 5, 1936, characterizing him as "one of the world's most distinguished entomologists and horticulturists." Several years after his death, Julian Nally wrote, "He was one of the greatest entomologists in the United States—this before he was twenty—and his collection of North American Butterflies was the third most complete in the world. He was widely read in the general field of literature with rare feeling and understanding for the field of philosophy. For fifty years he was one of Florida's most notable horticulturists."

As an expert entomologist, specializing in butterflies, of the many species he discovered, his legacy today rests with two that still carry his name—the brush-footed Mead's Wood-Nymph (*Cercyonis meadii*) and Mead's Sulphur (*Colias meadii*). His horticultural input helped W. H. Edwards successfully identify the feed and food plants for many North American butterflies, allowing the controlled egg-laying by pregnant females and the life-cycle study of individual species. These experiments led to confirmation that environmental factors controlled variations within butterfly species, in direct support of evolutionary thinking at that time.

Yet it was as a horticulturist that his influences were most far-reaching, and he made substantial contributions to the growth and development of ornamental horticulture in America and particularly to the Florida plant industry. Together with Henry Nehrling, he popularized and hybridized the caladium, and the industry settled in the Lake Placid area of Florida and flourished there. Today, all strap-leaved caladiums owe their origin to his efforts. He left his mark on citrus culture too, developing overhead water irrigation as a practical method of preventing citrus fruit damage from freezing temperatures.

His work with Nehrling led to the bi-colored Mead-strain amaryllis, which became the backbone of the amaryllis trade in America and a favorite of many of the gardens in the South. In later years, more and more gardeners started to replace the Mead-strain hybrids with the larger and more beautiful Dutch hybrids that were just as easy to grow and available with fragrance and in more shades and colors. But now long forgotten is that many white amaryllis growing in America today owe their origin directly to Mead's pioneering work.

His generosity in giving away plants and knowledge was legendary. In the early 1920s, he introduced Mulford Foster to bromeliads and the process of hybridization, encouraging him to popularize the bromeliad plant group to a larger audience, and in this process Foster became widely known.

35.1: C. x Meadii on the 1991 Singapore 30 cent stamp.

Aftermath & Legacy, 1936–1940

One of the earliest American hybridizers of the orchid, he created hundreds of new cultivars, almost all forgotten today. There was some recognition for his efforts many years later when *C x Meadii* was selected to illustrate the 30¢ Singapore stamp of 1991, as part of the International Orchid Conference held there. In addition, with his seminal work with Lewis Knudson, he prompted the start of the orchid industry with the discovery of sterile nutrient solutions for orchid seed germination.

In 1937, both Henry Nehrling and Theodore Mead became the first recipients of the Herbert Medal, awarded for their outstanding achievements in the advancement of knowledge of ornamental bulbous plants.

Before Mead's death, Jack Connery had once said to him in boyish enthusiasm, "Someday I am going to build a Memorial Garden for you," but Theo had laughed and modestly brushed the suggestion aside. When Connery enrolled at Rollins College in 1932 and met Edwin Grover, he found they both shared similar versions of this vision.

35.2: Mead Botanical Garden owes its existence to the vision and leadership of Jack Connery (left) and Edwin Grover. Derivative image created from original prints.

From an early age, Grover had a love of nature and there was a family interest in botany, focused through Grover's elder brother, Frederick, who became Professor of Botany at Oberlin College in Ohio where the College Herbarium was a national resource for botanists. Both brothers knew and admired Mead's work, and Edwin was of the opinion that every garden or public park in the State of Florida contained at least one plant that he had developed. Grover fully supported the idea of a Mead legacy garden that could also be used as a botanical resource for Rollins College, similar to his brother's in Ohio. So with Mead's passing in 1936, the stage was set for a perfect meeting of minds and dreams—for Connery to build a memorial garden to Mead, and for Grover to have access to a botanical garden for the benefit of students and staff of Rollins College. Before this could happen, however, there was a need to resolve the legal matter of who owned Mead's plants, and this fell to the will's executor, John Augustine Willis, to sort out.

Legally, the will Mead wrote in 1933 was clear—his orchids and greenhouse plants were to go to the Royal Palm Park at Paradise Keys. But also equally clear were the later letters he had sent to Jack Connery and Clifford Cole of Miami, promising the orchids divided equally between the two of them. The executor presented these facts to the representatives of the State Federation of Women's Clubs, owners of the Royal Palm Park, who in June 1936 magnanimously passed a resolution relinquishing all claims and giving the plants to the executor to decide on the best course of action. Following this, Cole gave his share to Connery, not wishing to split the collection and knowing of Connery's desire to establish a memorial garden.

Connery now had the challenging job of nursing the plants, and the orchid collection in particular, through the next few cold winters, while searching for suitable land for a botanical garden. He located what seemed to be a possible site, just six blocks from Rollins College between Pennsylvania Avenue and Maitland Avenue (now S. Denning Drive). The property lay in the southern part of Winter Park, bounded to the east by the tracks of the Dinky Line and extending into Orlando at Beverley Shores. Howell's Creek, connecting Lake Sue to Lake Virginia, ran through the site and provided flowing water to a number of ponds, a lower swamp and a large wetlands area. To the west, there was a sizeable region of upland hammock and pineland.

Connery shared his discovery with Grover and sought his help in acquiring the property. One by one the owners of the various parts of the property were approached and persuaded to donate their land. Money to clear the site and create trails and facilities came from the Work Progress Administration, the only WPA relief project in Central Florida. Ground was broken in January 1938 and two years later, on January 15, 1940, the Mead Botanical Garden opened its doors to hundreds of visitors, billing itself as "Florida's Finest Garden Spot." For many years after opening, it was the only recommended tourist attraction in the greater Orlando area, together with Sanlando Springs.

35.3: Hundreds of people attended the opening ceremony of Mead Botanical Garden in January 1940, held at the Orlando Beverley Shores entrance.

There would be no legacy to Theodore Mead today had it not been for the tireless, youthful energy and tenacity of Jack Connery on the one hand, and the influence and persuasive powers at the highest levels of Edwin Grover of Rollins College on the other. The checkered history of the Garden from the time it opened in 1940 until the present day remains to be told, but sufficient to describe it in later years, without Grover to defend it and keep the Barbarians from the gate, as not being worthy of the memory of Theodore Mead.

35.4: Tourist maps in the 1940s showed the Garden as one of only two major attractions in the greater Orlando area.

There were plenty of prominent people to recognize Theodore Mead's entomological and horticultural achievements immediately after his passing, but few of the tributes offered any insight into the man himself. He was a "man of high culture" who never sought publicity according to his friend Henry Nehrling, and this modesty probably explains his relative obscurity in the history of American horticulture. Unlike many other horticulturists at the time who tended to name every hybrid they created after themselves, he did not practice self-promotion, considering it a domain inhabited principally by the insecure. Probably the most accurate eulogy sums up the man as follows, "Mr. Mead was a gentleman of the old school, well educated, highly refined, widely traveled, of charming personality, but quiet and retiring in manner. He was a great lover of home life and a perfect host. His house was the Mecca for plant lovers, many of whom came from long distances to see his collections of rare plants." To this we can add generosity, in sharing his knowledge and freely giving away plant material to anyone that showed a real interest, and his paternal dedication to improving the moral and educational landscape of young people in Oviedo.

Despite all the accolades, there was a sense that when Theodore Mead looked back on his life's work in entomology and horticulture, he would never have considered any of it to be particularly significant, half expecting it to be destined for storage in a cobwebbed corner of some cupboard of history. Neither would he have been happy branded with the "wizard" moniker. In a characteristic show of undue modesty, he once downplayed the whole process of hybridization, saying, "They call a fellow a wizard if he takes the trouble to cross-pollinate a couple of blossoms. But as a matter of fact we do nothing more than the farmer who drops seeds in the furrow. We merely do systematically what the bees do for the farmer instinctively and haphazardly, and what the wind does because it cannot help it."

Digging around a little further, the concluding comments of his autobiographical article in the American Amaryllis Society Yearbook reveal what was important to him, "As I look back on four score years, the retrospect seems chiefly jeweled with happy friendships for young and old in all parts of the world. The things of the heart are the permanent ones in my life and nearest to what we creatures of a day may dream of as immortality." He once summed it up even more succinctly in the statement, "Love seems to be the only thing that is really worthwhile in the whole universe." Clearly, the real essence of life for this reluctant wizard was not insects and plants but friendships and love. So at journey's end, it is appropriate to finish this story of the life and times of Theodore Mead in the way he ended most of his letters to family and friends, with the simple adieu,

Acknowledgments

This book could not have been written without the kindness and tolerance of the custodians of the three primary sources of T. L. Mead material.

In the first place, I owe a special debt of thanks to Professor Wenxian Zhang and Darla Moore, who oversee the well-organized and extensive collection of Mead's letters in the archives of Rollins College, Winter Park. For more than five years, they provided generous support in allowing me access to study the collection. Secondly, a short walk away, at the Winter Park Public Library, the archivist Barbara White supplied friendly and professional help in the examination and copying of over a hundred of Mead's glass plate negatives in the History Center's Helen Connery Collection. Finally, my deepest gratitude goes to Cathy Raitt, who as a direct descendant of Webster and Anne Smith (née Edwards) opened up the treasure trove of family memorabilia at *Bellefleur*, the ancestral home of Mead's wife, Edith Edwards, in Coalburg, West Virginia. On the two visits I made there, in 2013 and 2015, she made me feel part of the extended Edwards family and did everything she could to assist me in telling Mead's story and that of her great-grandaunt Edith. Tom Willis, Betsy Rose, and Douglas Willis were other family descendants who gave or loaned me material and offered encouragement in this endeavor.

Many other friends and colleagues were of assistance with their own sources relating to Theodore Mead. Much information on the history of Mead Botanical Garden rests in The Winter Park Garden Club, and they were generous in granting access to their files, which included some late photographs of T. L. Mead. Kathleen Klare helped me more fully understand Mead's relationship with Henry Nehrling; Anne Michels of the Cathedral Church of St. Luke, Orlando, supplied me with information and images relevant to Mead's confirmation in 1903; and Howie Schaffer, Alumni President of Cornell Alpha Delta Phi, provided images of the 1878 brothers. Local expert historical lepidopterist John V. Calhoun, of the McGuire Center, Florida Museum of Natural History, University of Florida, shared many of his research findings of Mead and W. H. Edwards and made sure I got the butterfly side of Mead's life correct. To all these people, I acknowledge their help and involvement with sincere thanks.

The author, left, with Douglas Willis and Cathy Raitt, at Bellefleur *in June 2013, posing between the oil painting of Mary Mead on the wall and the Greek bronze vase on the floor, presented to Theo and Edith on their wedding day.*

Much gratitude goes to my friend and muse, Nancy Davila, who dispensed sustained encouragement, performed vigilant copyediting, and when I was faced with a blank page helped me see through the fog. Thanks, too, to Sue Foreman for her part in the copyediting process.

Mead was an early adopter of the new technology of photography, developing glass plate negatives and then hand coloring the resultant prints, many of which appear in this book. In keeping with this approach, a few originally black and white images have been sympathetically colored in Photoshop to bring them to life. For this skillful series of operations, I thank my youngest son, Elliott Butler.

Finally, financial support for publishing from Mead Botanical Garden, Inc. and the 2015 Smith Grant, administered by Rollins College and the Winter Park Library, is gratefully acknowledged.

Notes

Source Abbreviations – Primary Sources

Autobiography	"Theodore L. Mead, Naturalist, Entomologist and Plantsman—An Autobiography," *Yearbook of the American Amaryllis Society*, Vol. 2, 1935.
TLM	Theodore L. Mead.
RC-TLM	The Theodore L. Mead Collection, Archives & Special Collections, Rollins College, Winter Park, Florida. A vast collection of 31 boxes of letters and other printed material, some diaries and a few photographs.
WC	The Willis collection of private papers and photographs belonging to the Willis family, family descendants of Edith Mead, housed in the ancestral Edwards home at Coalburg, West Virginia.
WPPL	The Helen G. Connery Collection, Winter Park Public Library Archives, Winter Park, Florida.

Source Abbreviations – Secondary Sources

RC-HN	The Henry Nehrling Collection, Archives & Special Collections, Rollins College, Winter Park, Florida.
MG	Selected Mead letters from 1871 and 1872 in the McGuire Center for Lepidoptera & Biodiversity, Florida Museum of Natural History, Gainesville, Florida.
MAS-TLM	The Michael A. Spencer Collection on Theodore Mead, 1887–1939, Special Collections & University Archives, University of Central Florida, Orlando, Florida.
MAS-HN	The Michael A. Spencer Collection on Henry Nehrling, 1894–1997, Special Collections & University Archives, University of Central Florida, Orlando, Florida.
MAS-JN	The Michael A. Spencer Collection on Julian Nally, 1908–1977, Special Collections & University Archives, University of Central Florida, Orlando, Florida.
MAS-BR	The Michael A. Spencer Bromeliad Research Collection, 1754–2003, Special Collections & University Archives, University of Central Florida, Orlando, Florida.

CHAPTER 1: When Samuel Met Mary

The description of Theo's birth on February 23, 1852, is contained in Mary Mead's diary, RC-TLM.

Theo's grandfather, Ralph Mead, was the owner of a wholesale grocery business at 13/15 Coenties Slip, close to the East River and one of the dominant importing houses of the day. In 1837/8, he had three elegant red brick and brownstone row houses, number 108, 110 and 112, constructed on Second Avenue near Seventh Street. Today, the only surviving structure is No. 108, now renumbered 110, and known as the Isaac T. Hopper Home. Source: New York City Landmarks Preservation Commission, October 13, 2009, Designation List 419 LP-2331, "Ralph and Ann E. Van Wyck Mead House (later Isaac T. Hopper Home of the Women's Prison Association), 110 Second Avenue, Manhattan," http://www.nyc.gov/html/lpc/downloads/pdf/reports/mead.pdf, [accessed 20 December 2015].

The extract recalling their first meeting comes from a letter, dated December 29, 1849, sent from Mary Mead to Samuel Mead, RC-TLM.

The Luqueers were descended from a Huguenot family named "L'Escuyer", pronounced "Lay Qwee Ay." The origin of the name in old French means "squire" or "shield bearer." Jan L'Escuyer was born in Paris in 1635 and came to America in 1658 to escape religious persecution, sailing from Amsterdam with Dutch colonists on the ship De Bruynvis (The Brownfish).

At school, Mary Luqueer was an exemplary student and there are many examples of her diligence in school papers and certificates of achievement in RC-TLM, Box 17.

Women made up the majority of converts to the Awakening, and mothers were seen as the moral and spiritual foundation of the family, tasked with actively instructing their children in matters of religion and ethics. The phrase "drenched in wickedness" was used in evangelical circles to portray the world as full of sin and to justify its elimination by social activism, as quoted in http://www.worldlibrary.org/articles/evangelicalism, [accessed 20 December 2015].

The statement regarding Sarah Mead's struggle with one of her children's religious orientation is from the book *Our Excellent Women of the Methodist Church in England and America* (New York: James Miller, 1873), 230.

Theo's father left Wesleyan University in his junior year and became a "gentleman of elegant leisure," according to the *Alumni Records of Wesleyan University*, 3rd ed. (Hartford: Case, Lockwood & Brainard, 1883), 375.

William Mead was born in Lydd in Kent, England, and came to America in April 1635 as an economic immigrant with his two sons, Joseph and John. Tired of organized religions and their harsh doctrines and church control into their everyday lives, they settled in Connecticut around Greenwich in 1641.

Samuel Mead wrote to Mary on June 27, 1844 describing his dream of living in the country and experiencing "the breath of flowers and the perfume of hay," RC-TLM.

Mary's concerns that they might be religiously mismatched and she would be happier if he were a minister, and Samuel's reply that he would make a bad preacher but welcomed the love of a guardian angel, come from letters in RC-TLM. Both letters are undated drafts, presumably written 1844.

"I thank my Heavenly Father for the love of one who cares for my soul," is from Mary Luqueer to Samuel Mead, August 24, 1844, RC-TLM.

Details of their marriage appears in *US Dutch Reformed Church Records in Selected States, 1639–2000*, New York, New York Marriages, Book 32, 140 (www.ancestry.com), and James Hardie, *The Description of the City of New York* (New York: Samuel Marks, 1827), 175.

Tragically at the age of 51, Samuel's mother, Sarah Holmes Mead, contracted dysentery in October 1842 and died after a three-week illness. In the following year, Ralph Mead married his second wife, Ann E. Van Wyck, a daughter of

General Abraham Van Wyck, of Fishkill, New York, and acquired by marriage land and farm property there.

CHAPTER 2: Hudson Valley Upbringing

Expenses and income from the Fishkill farm are documented in notebooks from the 1850s. Other ledgers and notebooks of the time (RC-TLM, Box 16) show how carefully the Mead family tracked expenses.

Theodore Mead's interest in entomology and horticulture started when he was very young. The "crowning glory" of a large amaryllis and his nursemaid throwing a caterpillar of the *Cecropia* Moth into the fire are remembered in "Theodore Luqueer Mead—The Making of a Plant Breeder," *The Florida Trucker*, Vol. 2, No. 6, 25 June 1925.

Mary Mead's diary for 1853 in RC-TLM contains entries related to disciplining the children, and examples of Samuel's poetry written to her.

With Mary away, Samuel took on the role of teacher and reported on Theo's progress in a letter dated May 16, 1859. Her chief worry was whether standards in religious indoctrination were being maintained, as she inquired in a letter sent March 22, 1860, (both letters, RC-TLM).

Theo admitted to his father's indulgence in a letter to one of his pen friends, Clarence Gilbert, dated October 27, 1915, RC-TLM. On Sundays, he was allowed to read educational texts, like books on chemistry, instead of the Bible (Mead, *Florida Trucker*, 4).

References to old Diddy the cat following Theo around are from a letter dated June 9, 1857, Samuel Mead to Sammy, RC-TLM. Further insights into Theo's childhood and his lack of social contact with playmates his own age are contained in Mead, *Autobiography*, 3–14, and Mead, *Florida Trucker*, 4.

The letter from Mary Mead to her eldest son Sammy over his poor schooling performance, and his father's response was sent April 2, 1860, RC-TLM. Mary was summoned to New York by the news that her father had taken a turn for the worse. He died on April 25, 1860, aged 67.

Elie Charlier (1826–1896), the son of a French Protestant clergyman, founded the school that carried his name as a French and English School for boys in New York. "We must train the best colt when young," is taken from Elie Charlier to Samuel Mead, October 12, 1860, RC-TLM.

The start of the American Civil War was a likely contributory factor in the Mead family's decision to go to Europe. Pressures to enlist in the Union cause were growing by July 1861 and conscription was round the corner. Theo's father had completed seven years of cadet military training with the First Division, First Brigade, of the Ninth Regiment of the New York State Artillery, granting him exemption from future military and jury service (certificate in RC-TLM).

CHAPTER 3: Schooling in Germany & America

On December 17, 1861, Samuel Mead wrote to Mary Mead in New York from L'Hotel du Louvre in Paris, telling her of their son's growing maturity and social skills, RC-TLM. Mention of his improving mastery of the French language are in letters sent February 13 and 28, 1862, RC-TLM.

Elie Charlier wrote to Mary Mead on May 13, 1862, RC-TLM, praising Theo's diligent attitude to learning and regretting the gymnasium accident.

The Mead's lodging house address in Heidelberg comes from an envelope sent to Mary Mead from her sister Lottie, dated March 18, 1863, RC-TLM.

The spa industry began in Homburg with the discovery of Elizabeth's well (Elisabethenbrunnen) in 1834. John MacPherson wrote a popular reference guide to spas, *The Baths and Wells of Europe* (London: Edward Stanford, 1888), praising Homburg's "good natured girls."

The Meads enjoyed listening to classical music in the grounds of the Kursaal, public halls used by visitors at watering places in Germany. Details of these events are from Mary Mead's diary for 1863, RC-TLM.

The quotations concerning Theo's time at school in Frankfurt are taken from a single sheet undated manuscript, probably a draft of part of a letter he wrote,

RC-TLM. The observation that practically all the boys at the school were ardent collectors comes from *The Florida Trucker* article. Theo's self-confidence when it came to traveling comes from an undated extract of a letter from his father to a friend, RC-TLM.

When his parents moved to Homburg, they chose lodgings in the small village of Gonzenheim, just outside the town. The Homburg Railway, one of the first railways in Germany opened in 1860, aimed at increasing patronage to the spa and casino.

Mary Mead's sermonizing to her boys was well meant but consisted of endless repetition about religious piosities; see Boxes 12 and 13, RC-TLM. It has not been possible to find her reply to Theo's argument that if he must go with the other boys to church he would also like to go with them to drink beer, TLM to parents, June 27, 1864, RC-TLM.

Theo was a precocious child and quickly became bored with subjects he thought a waste of time. The extract is from a letter to his mother, September 24, 1863, RC-TLM. The New York passenger lists for ships arriving in New York on August 1864 can be found at ancestry.com.

The reference to teaching in the New York school being largely "parroting recitation by rote," and the use of caning, appear in Mead, *Autobiography*, 3. Tammany Hall, or simply Tammany, was the name given to a powerful political machine that ran New York City throughout much of the 19th century. "No organized athletics" is a comment in Mead, *Florida Trucker*, 4.

The report of the death of Ralph Mead is recorded in *Commemorative Biographical Record of Fairfield County, Connecticut* (Chicago: J. H. Beers & Co., 1899), 361.

Theo's class position in Science and the circumstances leading to his European tour with his mother come from Mead, *Autobiography*, 3. Theo was indulged on this and many other occasions, getting his way with things so long as they were purchasable. It would be a hard lesson for him to learn later on in life that there were many things in life that money could not buy.

CHAPTER 4: The Grand European Tour

Theo wrote to his brother Sammy on March 27, 1867, RC-TLM, while onboard the SS *Fulton* off the Irish Coast, telling him about the rough crossing and his bout of seasickness.

The site chosen for the Exposition Universelle of 1867 was the Champ de Mars, the great military parade ground of Paris.

In Rome, Theo admitted in a letter sent to Sammy on April 25, 1867, RC-TLM, that he was lonely for male company his own age. His mother's attempts to speak Italian are recorded in the same letter. Theo described the events of Holy Week in Rome as "grand humbug," in TLM to father, Hotel Minerva, Rome, April 21, 1867, RC-TLM.

Theo's initial impressions of Venice are contained in a letter dated June 12, 1867, to his father and Sammy, RC-TLM. The description of the famous stalactite cave at Adelsberg being like an upside down Milan cathedral is from "The Cave of Adelsberg; An Underground Wonder of Southern Austria," *New York Times*, 28 Mar. 1881, 1.

Theo's admiration for St Petersburg and its magnificent buildings is expressed in a letter to his father and brother, dated July 14, 1867, RC-TLM. The massive palm he saw in the Botanical Garden is mentioned in his autobiography.

A letter from Theo to his father and brother sent from Berlin on August 29, 1867, RC-TLM, tells of meeting many friends from their previous time in Germany. For Theo, the meeting with Otto Staudinger in Dresden was transformative. Staudinger was an entomologist and natural history dealer, specializing in the collection and sale of insects to museums, scientific institutions, and individuals. The butterfly collection his mother bought for Theo was extensive, costing $50 or approximately $800 in 2015 terms. According to Mead's autobiography, this event marked the start of serious entomological collecting.

The report of baggage examination at Dover appears in a letter to his father dated

September 26, 1867, RC-TLM. Their passage home from Liverpool was on the Cunard single screw SS *Tripoli*, launched for the Cunard Mediterranean service.

Theo's overall analysis of the European trip as having a heavy religious bias in an attempt to improve his mind, is contained in *The Florida Trucker* article.

CHAPTER 5: First Visit to Florida

Theo's election at age sixteen to the American Entomological Society comes from his autobiography.

In 1869, the favored way of getting from New York to Florida was by steamer to Charleston or Jacksonville, connecting there to ships of the Charleston and Florida Steamship Company for the journey up the St. Johns River; see *The Guide to Florida – The Land of Flowers* (New York: Catlin & Lydecker, 1874), 19. Captain Brock's steamer *Darlington* left Jacksonville at 8:00 a.m. on Saturday, timed to receive passengers discharging from ocean-going ships, arriving on Sunday in Palatka, from which it departed at 5:00 a.m. on Monday morning, docking at Enterprise that evening adjacent to the conveniently located *Brock House*. The $9 fare from Jacksonville comes from Floyd and Marion Rinhart, *Victorian Florida: America's Last Frontier* (Atlanta: Peachtree Publishers Limited, 1986), 56.

The perils of snakes slithering down the hanging Spanish Moss onto the decks of the St. Johns steamboats is noted at Virginia Cowart, "Paddlewheelers on the St. Johns," 2005, http://www.cowart.info/Monthly%20Features/Paddlewheel/Paddlewheelers.htm, [accessed 1 November 2015]. The attraction of alligator teeth as tourist ornaments is reported in Edward A. Mueller, *Steamboating on the St. Johns*, (Melbourne, FL: South Brevard Historical Society, 1980), 30.

The quotation denouncing the practice of shooting wildlife from the decks of St. Johns steamers is originally from a short article by Harriet Beecher Stowe, "A Southern Snow-Storm," *The Christian Union*, VII:16 (1873): 302. It can also be found in John T. Foster Jr., *Calling Yankees to Florida - Harriet Beecher Stowe's Forgotten Tourist Articles* (Cocoa, FL: The Florida Historical Society Press, 2012), 133. The Mead boys' own experiences shooting alligators are from Sammy's diary of March 15, 1869, RC-TLM.

The reference to the wingspan of the shot osprey, and the description of the trips up to Lake Harney on the steamer *Hattie*, a shallow draft vessel owned by Captain Brock and named after one of his daughters, are from a letter Sammy sent to a friend from Enterprise dated April 1, 1869, RC-TLM. The "openness when they smile" comment appears in a letter Samuel sent from New York on March 29, 1869, RC-TLM. A complete account of the adventures of his trip was subsequently written by Theo and appeared as "Florida and Alligators" in the April 1870 issue of *The Microcosm* published by the Poughkeepsie Military Institute (Poughkeepsie, N.Y.); copy in RC-TLM.

Papilio calverleyi, at the time thought to be a rare new species was later discovered to be an extreme aberration of the subspecies *Papilio polyxenes asterius* – the Black Swallowtail. This specimen is with the rest of the T. L. Mead butterfly collection in the Lepidoptera section of the Carnegie Museum of Natural History at Pittsburgh.

CHAPTER 6: With W. H. Edwards in West Virginia

The invitation from W. H. Edwards to come to West Virginia to collect butterflies is mentioned in Mead, *Autobiography*, 4.

The collection of sources used in describing W. H. Edwards' early collecting life and move to Coalburg as a coal mine operator are: "Brief Biographies – William Henry Edwards," *The Lepidopterists' News,* Vol. 1, no. 1, May 1947, 8; William H. Edwards, *A Voyage up the River Amazon* (London: John Murray, 1861); W. S. Laidley, *History of Charleston and Kanawha County, West Virginia,* (Chicago: Richmond-Arnold Publishing Co., 1911), 316; and Cyril F. dos Passos and William Henry Edwards, "The Entomological Reminiscences of William Henry Edwards," *Journal of the New York Entomological Society*, Vol. 59, No. 3, (Sep., 1951), 129–186.

W. H. Edwards was a direct descendant of Jonathan Edwards, America's foremost theologian. Despite this, he became a Darwinian and agnostic. The account of this journey is covered in William Leach, *Butterfly People,* (New York: Pantheon Books, 2013), 8–13.

The second half of the nineteenth century coincided with a period when the USA battled for supremacy with Europe to become a world-class astronomical power. Sammy applied his telescope to the serious study of interstellar objects and joined the debate over the Million Dollar Telescope, becoming somewhat of an expert on the subject—see, for example, letter to Editor from S. H. Mead Jr., "Mr. Richard A. Proctor and the Million Dollar Telescope," *Scientific American*, 7 March, 1874, 148. The story of this race for global astronomical supremacy is recounted in Trudy E. Bell, "The Great Telescope Race," *Sky & Telescope*, 121, Issue 6, (2011), 28–33.

The US patent for Sammy's explosive bullet is "Improvement in explosive bullets," Samuel H. Mead Jr., of New York, No. 133,714, Granted December 10, 1872.

The tintype photographic process, popular during the 1860s and 1870s, produced a mirror image of the subject as a positive silver image in a collodion binder on an opaque, non-reflective support surface, usually a thin sheet of lacquered iron. At this stage in the development of photography, there were neither camera shutters nor f-stops, and exposures were made by removing the lens cap for a second or two and then replacing it.

In May 1873, the house number on Madison Avenue changed from 596 to 674; see letter TLM to Will Edwards, May 2, 1873, RC-TLM.

Adult butterflies take nectar from a range of flowering feed plants but generally these are not the plants where eggs are laid, since caterpillars are selective about which food plant they consume. Theo's ability through his horticultural knowledge to discover the food plants of butterfly species was an important factor in Edwards' future egg-breeding program, as recognized in Edwards' "Entomological Reminiscences," 142. The systemic breeding program and the importance of the C&O flag stop at Coalburg in providing eggs from other parts of the world are detailed on page 146 of the same reference, and in Leach, *Butterfly People*, 19–20 & 133.

CHAPTER 7: Chasing Butterflies in Colorado

The Wheeler Surveys, expeditions from 1869 to 1874 led by First Lieutenant George Montague Wheeler, were initiated to map and characterize areas of the United States lying west of the 100th meridian. Their principal purpose was to create accurate topographic maps of the region, but additionally to explore and record mineral resources, geology, vegetation, water sources, zoology and agricultural potential. Although supervised by the military, they included eminent scientists of the day, and Mr. W. H. Edwards was one of the entomological experts.

Sending specimens directly back to Edwards in Coalburg turned out to be a fortuitous decision. Many of the other specimens brought back from the Wheeler exploration that year were stored in Chicago, and lost in the Great Chicago Fire of October 1871.

Meeting Buffalo Bill at the American House in Denver and turning down his offer of traveling together is reported in a letter Sammy sent to his parents dated June 2, 1871, RC-TLM. Bonner's Ledger was a weekly New York story paper.

The source for the quotations that the Indians were friendly and only killed one man last week, and that Theo was tired of eating trout, is F. Martin Brown, *Chasing Butterflies in the Colorado Rockies with Theodore Mead in 1871,* Bulletin No. 3, Pikes Peak Research Station, Colorado Outdoor Education Center, Florissant, CO, 1996, 4 & 33. The unwanted attention of bedbugs is noted on page 38 and the account of their ascent of Mt. Lincoln on page 51.

The butterflies taken each month in Colorado were reported by Theo in Correspondence, "Insects and Flowers in Colorado," *Popular Science Monthly*, Volume 11, May 1877, 104. The numbers are huge by today's standards, and a reminder of how human progress and development can affect previously pristine environments.

The account of Theo's visit to the Florissant fossil beds is given in Brown, *Chasing Butterflies*, 62. Scudder's research paper appeared as Samuel Hubbard Scudder,

"Fossil Coleoptera from the Rock Mountain Tertiaries," *U.S. Department of the Interior, United States Geological and Geographical Survey of the Territories, Volume 2, Number 1,* (Washington, DC: GPO, 1876), 77–78. Theo was stated to be the first person to recognize the scientific importance of the site in F. Martin Brown, "A Note about Florissant Fossil Insects," *Entomological News* 92, (1981), 165–166. The site was named after Florissant, meaning "blossomy" in French, a settlement close to the site founded in 1872 by James Castello, and named after his hometown of Florissant, Missouri.

The reference to "pictured saints and holy families" relates to his 1867 European tour with his mother when he was taken, at times unwillingly, from one religious-oriented location to another. The source is a letter from TLM to George M. Dodge, 31 July 1871, quoted in Brown, *Chasing Butterflies*, 41. The Meads collecting time in Panama on the way back to New York is described on page 73 of this volume.

The first reconstruction of Theo's Colorado itinerary appeared as F. Martin Brown, "Itineraries of the Wheeler Survey Naturalists 1871 – Theodore L. Mead," *The Lepidopterists' News*, Vol. 9 (6), 1955. The publication *Chasing Butterflies in the Colorado Rockies with Theodore Mead in 1871* followed in 1996. Subsequent research by John Calhoun has added much more detail and accuracy to his collecting time there; see John V. Calhoun, "The correct publication date of the report upon the collections of diurnal Lepidoptera made in portions of Colorado, Utah, New Mexico, and Arizona by Theodore L. Mead," *News of the Lepidopterists' Society,* 55(3), 2013, 96–99, and John V. Calhoun, "An updated itinerary of Theodore L. Mead in Colorado in 1871 with type locality clarifications and a lectotype designation for *Melitaea eurytion* Mead (Nymphalidae)," *Journal of the Lepidopterists' Society,* 69(1), 2014, 1–38.

CHAPTER 8: New Species or Darwinian Variant?

After two years' work, Theo wrote up the discoveries of the various Wheeler expeditions with the help of W. H. Edwards. The reference is: Theodore L. Mead, "Report upon the Collections of Diurnal Lepidoptera made in portions of Colorado, Utah, New Mexico, and Arizona, during the years 1871, 1872, 1873, and 1874," Chapter VIII, 735–793, in "Report upon Geographical and Geological Explorations and Surveys west of the One Hundredth Meridian," Chapter 1, Volume V - Zoology, (Washington, DC: GPO, 1875).

On 21 February 1871, the original Edwards' family house and contents burned down, although the butterfly material was saved. A description of this event is in John V. Calhoun, "The Extraordinary Story of an Artistic and Scientific Masterpiece: *The Butterflies of North America* by William Henry Edwards, 1868–1897," *Journal of the Lepidopterists' Society*, 67(2), 2013, 78. There are no records of the house name as *Bellefleur* before around 1909. In 1990, the house was placed on the National Register of Historic Places (Edwards, William H. & William S. House, SR 61 NE of Cabin Creek, Coalburg, Kanawha County, West Virginia).

Louis Agassiz was a Harvard University professor of zoology and geology. He was an influential and outspoken critic of natural selection as proposed by Charles Darwin. He never altered this view despite Edwards' egg-breeding program showing three variants of polymorphism in a single species, the zebra swallowtail, see Leach, *Butterfly People*, 23 & 157. The short lifespan of butterflies and moths allowed such evolutionary processes to be directly observed.

Edwards' volumes on *The Butterflies of North America* were judged as important as Audubon's books on birds by Grinnell (Fordyce Grinnell, "The Work of W. H. Edwards," *The Lepidopterist,* Volume 1, 1917, 92) and, by many, the best books written on American butterflies (Leach, *Butterfly People*, 134).

The warm atmosphere in the Edwards home was commented on in TLM to Sammy, July 11, 1872, MG.

Mary Mead's religious sermonizing and nagging affected both of her sons, as shown in the letter from Sammy to Theo, dated April 7, 1875, RC-TLM.

Theo's low threshold for boredom was one of his characteristics, and it seemed that the moment he achieved prominence in one area he wanted to move on to another. In the summer of 1873, now a butterfly expert to rival W. H. Edwards, he wrote to his father, "I collect a little, read, swim, and loaf around generally. I am really becoming very lazy in the matter of collecting; as there are no very particular rarities in the neighborhood." (Letter, dated August 11, 1873, RC-TLM).

Theo's response to the challenge Will Edwards set him about choosing his future profession is contained in a letter dated April 6, 1873, RC-TLM. His admission that civil engineering was the wrong choice at Cornell appears in his autobiography.

CHAPTER 9: Secret Societies & the Delta Upsilon Fraternity

Theo's first impressions of the Cornell campus at Ithaca are from letter to mother, October 1, 1873, RC-TLM.

The row between the independent and the Greek fraternities is covered in "Delta Upsilon One Hundred Years 1834–1934," published by the fraternity in 1934, 122. The acerbic written reaction to the founding of the Delta Upsilon fraternity at Cornell appeared in *The Cornell Era*, Vol 1, No 22, May 22, 1869, 4.

"The Sad Tale of Mortimer Leggett," and the soul-searching inquest after this accidental death by hazing, is told on page six of *The Cornell Daily Sun*, Volume XCVII, Number 108, 20 March 1981.

Theo told his father of his official election to the Delta Upsilon Society on March 8, 1874, RC-TLM. His 1875 essay on "Evolution of Sense-Organs," was published in *The Cornell Review*, February 1875, 236–238. It evoked a predictable response from his mother, February 19, 1875, RC-TLM.

The description of a typical Delta Upsilon social evening, singing college songs and drinking lemonade through straws, comes from a letter TLM sent to Sammy, July 1, 1874, RC-TLM.

The circumstances surrounding Sammy's death have been put together from the following sources: letter father to TLM, May 21, 1876, RC-TLM; "Fatal Accident to an Inventor," *New York Daily Tribune*, Friday, May 21, 1875; and Mary Mead's diary entry of May 20, 1875, RC-TLM. Theo's personal lament is from a letter he sent to *Scientific American* on June 9, 1875; the formal obituary to Sammy appeared in the issue of June 26, 1875.

His father's meditative visit to Green-Wood cemetery on the anniversary of Sammy's death is noted in letter dated May 21, 1876, RC-TLM.

CHAPTER 10: Fraternity Conflicts

Beardsley & Mackey, Cascadilla Art Gallery, 7 Linn Street, Ithaca was a favorite studio for portrait photography. Theo's delight with the results of retouching and the Beardsley testimonial are from a letter to his parents dated March 10, 1876, MG.

The news of Theo's forty-one varieties of cacti and his first hybridizing efforts are from letters to his father dated September 21, 1875, RC-TLM and November 22, 1876, MG.

Theo enjoyed being at the center of things and he needed no encouragement to take charge of the financial matters of Delta Upsilon to give him practice in business affairs. He talked of this and his new role in a letter to his parents dated May 7, 1876, RC-TLM.

The tensions within Delta Upsilon and the description of the events leading to the expulsion of Edwards and Theo's resignation are from various sources: William Sheafe Chase, "The Delta Upsilon Quinquennial Catalogue," published by the fraternity, 503; letters sent by TLM to Eugene Frayer in New York, March 19, 1887, RC-TLM; and Charles Harmon in Chicago, undated but presumed 1877, RC-TLM. Frayer and Harmon were ex-Delta Upsilon senior figures who supported Theo in his early days at the fraternity.

The Cornell University records show that Theo formally attended from 1871–1879, completing 13 academic terms (trimester system) and receiving his first degree, a Bachelor of Civil Engineering in 1877. From 1878 to 1879, he completed several terms of post-graduate work in natural history. His Alpha Delta Phi fraternity records him as "class of '77" (first degree) and he appears in the fraternity group photographs of 1878 and June 1879.

Theo considered the day of his initiation into the Alpha Delta Phi fraternity as one of the most memorable of his life (Mead, *Autobiography*, 5). In a letter to Eugene Frayer, dated April 27, 1877, RC-TLM, he recognized that "I now belong to a society I should have joined when a freshman."

His father's reply to him about feeling "intensely unhappy" is dated March 19, 1877, RC-TLM. The advice to "dodge all botheration" comes from a letter his father sent on May 20, 1877, RC-TLM.

CHAPTER 11: Escaping Botheration out West

The sea journey up the coast from Panama, and their time in Acapulco, are made known in Mead's autobiography and in the diaries of his father for 1878 (RC-TLM). A letter from TLM to W. H. Edwards on April 28, 1878, RC-TLM, tells how good it was to eat decent food again.

The Southern Pacific Railroad had recently opened a route from San Francisco to Los Angeles, 500 miles away and was offering an introductory fare of only $10. Theo's description of Los Angeles as not much more than a village is from his autobiography, as is the decision to move inland to stay at Cogswell's Sierra Madre Villa. This lodging is described in Ben C. Truman's *"Tourists' Illustrated Guide of California,"* (San Francisco: Crocker & Co., 1883), 89–90, and as Southern California's first famous resort, in http://eastofallen.blogspot.com/2008/06/sierra-madre-villa.html, [accessed 25 November 2015].

The story of Theo's missed opportunity to buy real estate in Los Angeles appears in Mead, *Autobiography*, 7.

Elias Jackson Baldwin's lucky moment came when he received 2,000 shares of the Ophir Mining Company, then worth a few cents a share, in payment of a debt. With the discovery of the Comstock Lode in Nevada in 1859, the shares jumped to several hundred dollars each. Baldwin invested his wealth in land, owning thousands of acres in southern California where the communities of Arcadia, Sierra Madre, and Monrovia are now located (https://en.wikipedia.org/wiki/Lucky_Baldwin, accessed 25 November 2015). A description of the Mead's visit and Theo's comment about the scale at which Californians did things are from a letter he wrote to W. H. Edwards from the villa on May 14, 1878, RC-TLM.

Snow's Hotel, *La Casa Nevada*, was located below the granite dome, Liberty Cap, and at the west end of Little Yosemite Valley, situated within the spray of Nevada Fall at an elevation of 5,360 feet. Albert and Emily Snow operated this pioneer hotel for almost 20 years before it closed in 1891. Yosemite innkeepers, like the Snows, provided a valuable service to the increasing number of tourists visiting the Valley in the late nineteenth century. Details including the "ain't that eleven?" quote come from Hank Johnston, "Yosemite's La Casa Nevada (The Snow House)," *Yosemite*, Winter 2004, Vol 66, No 1, 3–5.

The description of their time in Yosemite and the "Every day seemed more wonderful" quote are from Mead, *Autobiography*, 8. The Glacier Point expeditions are recorded in letters to W. H. Edwards, dated June 11 and June 20, 1878, RC-TLM.

Details of the journey north from Yosemite to Tallac Point and into Nevada, Utah and Wyoming and the capture of *Chionobas ivallda* come from his father diary entries for July, August and September 1878, RC-TLM. A letter to W. H. Edwards dated July 8, 1878, tells of the shipment of 1,500 butterflies (RC-TLM). A communication from Theo to Will dated August 6, 1878, RC-TLM speaks of a "great store of prey" taken in the way of butterflies from the Lake Tahoe/Donner Pass region and it is likely that one of these was later named *Gaeides* (now *Lycaena*) *editha*, in honor of W. H. Edwards' eldest daughter. Source: Pyle, R. M. 1981, *National Audubon Society Field Guide to North American Butterflies* (Alfred A. Knopf, Inc., New York, New York, USA, 21st. printing, 2010), 409.

The letter from his father detailing the expenses of the trip at around $3,000 (equivalent to approximately $70,000 in 2015 terms), expressing the view that the fun was worth it, and reminding him that he still had $50 or so in his pocket to get back to New York, was sent on September 23, 1878, RC-TLM.

CHAPTER 12: Alpha Delta Phi Brotherhood

The history of the building of the first Alpha Delta Phi Chapter House is from "A Comprehensive History of Alpha Delta Phi," by Marc B. Zawel '04, https://www.adphicornell.org/public6.asp, [accessed 28 January 2016]. Theo's mother would continue all her life to persuade him to convert to her form of evangelical religion, trying every trick in the book, including this one, to bring it about. The "Don't do it, Ted" quote appears in Mead, *Autobiography*, 5.

A letter to his parents lists the coursework at Cornell he chose for his post-graduate studies in Botany (October 4, 1877, RC-TLM).

Theo refers to many of his fraternity brothers in letters to his parents (RC-TLM), and there is a full list in "Catalogue of the Alpha Delta Phi 1832–1915," (New York: Alpha Delta Phi Fraternity, 1915), 261–263.

Edward House's extensive vocabulary is commented on in Mead's autobiography, and no doubt stood him in good stead for his later role as "second personality" to Woodrow Wilson, see Charles Seymour, "The Intimate Papers of Colonel House," (Boston & New York: Houghton Mifflin Co., 1926), 114. Harry Robie's father was Rear Admiral Edward Dunham Robie (1831–1910), who had a distinguished naval career as an inventor and marine engineer.

Concern about Theo's future led Will to suggest an academic career at Harvard University under Professor Hermann Hagen—TLM to father, May 27, 1879, RC-TLM. Theo was tempted, but his father suggested Columbia Law School (Mead, *Autobiography*, 8).

News of his position as a stockbroker is from *The Cornell Era* Vol 13, January 14, 1881, 167.

Theo left Cornell considerably wiser, proof that, as Will Edwards predicted it would take a whole lot of "calf" out of him (Mead, *Florida Trucker*, 4).

Theo's brief expedition to Newfoundland with his father over the summer of 1880 and their encounter with the black flies is recounted in Mead, *Autobiography*, 8.

CHAPTER 13: Florida Land Purchases

Representative examples of Florida tourist promotion in the later part of the nineteenth century are: Foster, *Calling Yankees to Florida*, 17; *The Guide to Florida – The Land of Flowers* (New York: Catlin & Lydecker, 1874), 19; and H. K. Ingram, *Florida Tourists and Settlers Guide to Florida* (Jacksonville: Da Costa Publishing, 1895-96), where the verse "It is Florida" appears in the preface.

"The orange is surer than gold" is taken from Mark Andrews, "1880s Writer Sees Gems In Winter's Flowers And Gold In Florida's Orange," *Orlando Sentinel* 15 May 1994, http://articles.orlandosentinel.com/1994-05-15/news/9405130692_1_newell-orange-county-florida, [accessed 28 January 2016].

Newspapers adding to the promotional imagery come from Bill Bond, "Early Weeklies Offer Look into Lake's Past," *Orlando Sentinel* 23 May 1987, http://articles.orlandosentinel.com/1987-04-23/news/0120330133_1_sumter-county-weekly-newspapers-eustis, [accessed 28 January 2016].

John Macdonald's role in helping to grow the Eustis region is covered in Ormund Powers, "Local Pioneer A Force Behind State's Growth," *Orlando Sentinel*, 11 March 1998, http://articles.orlandosentinel.com/1998-03-11/news/9803110452_1_eustis-mount-dora-macdonald, [accessed 8 November 2015]. Eustis was in Orange County until 1887 when it became part of Lake County.

Details of the Mead's arrival in Florida come from his father's diary entries, June/July 1881, RC-TLM. "Holding on with the tenacity of alligators" and using kerosene against flies come from his diary entries of July 1 and August 1 respectively.

An advantage of the Eustis region at that time was the rail connection to the North to Astor and the St. Johns River steamers. A. S. Pendry homesteaded in Eustis in 1876, started growing citrus, and opened the Ocklawaha Hotel in 1877, naming the settlement after himself.

His father's diary entry of July 7, 1881, RC-TLM, records the purchase of their first Florida property.

Before designation of eventual ownership, particularly for rural, wild or undeveloped land, the United States used a rectangular surveying system to spatially identify land parcels. Section-Township-Range (STR) units became a standard reference system for land transfers and official documents.

His father's diary entry for August 13 and 24, 1881, RC-TLM, records the fruit tree and pineapple plantings. In a letter to parents, December 6, 1881, RC-TLM, Theo proposed to buy the new family sailboat manufactured by the Racine Boat Company, of Racine, Wisconsin. The Crescent Grove Experimental Farm letterhead is from RC-TLM.

By 1890, Theo and Edith had relocated to Lake Charm but his parents spent their winters on the Eustis property until 1899. In 1890, the land was platted to resolve a boundary dispute at the shoreline of the lake between the Meads and the adjoining landowner, Colonel Lane. The plat shows the site of the Mead home on the bluff with views of Lake Eustis, and the cultivated and orange grove areas; see www.officialrecords.lakecountyclerk.org, plat book 1, page 6, 6/2/1890. In the early 1900s the property was sold and by 1940 it was being promoted as the first planned community in Eustis, designed in a circular pattern like Washington, DC—Ruth Downs Akright and Betty Slaven McClellan, *Eustis* (Mount Pleasant, SC: Arcadia Publishing, 2011), 111.

CHAPTER 14: Marriage to Edith Edwards

Edwards' time in England to secure capital for coalfield development comes from Otis K. Rice and Stephen W. Brown, *West Virginia: A History*, 2nd ed. (Lexington: University of Kentucky Press, 1993), 86. Details of Edith's baptism and return to the USA are from ancestry.com.

Letters to Edith's Aunt Sarah (December 14, 1877, and February 1, 1878, WC) from the Brooklyn Homeopathic Hospital, 109 Cumberland Street, speak of Edith's time there as a nurse. Other details are from "Mrs. T. L. Mead Dies at Home near Oviedo," *Orlando Morning Sentinel*, October 26, 1927. A letter from TLM to his mother, dated September 8, 1873, RC-TLM, claims his one-tenth ownership of the organ.

An example of Edith's skill as an artist depicting butterfly chrysalises can be found in Samuel Hubbard Scudder, *The Butterflies of the Eastern United States and Canada, Volume 3* (Cambridge: Scudder, 1889), Plate 84, Figure 4, page 2300.

Will's Christmas invitation to TLM came in a letter dated December 15, 1881, RC-TLM.

It is hard to put a value on the combined Mead/Luqueer fortune with an asset base of cash, stocks, and desirable Manhattan properties, but several million dollars at today's evaluation would seem reasonable. It was a handy cushion but Theo was determined to pay his way through citrus growing.

Edith Edwards was directly descended from the revivalist preacher Jonathan Edwards, author of the famous sermon *Sinners in the Hands of an Angry God*, delivered to his own congregation in Northampton, Massachusetts to unknown but presumably electrifying effect. Theo's concern as to whether Edith had inherited such revivalist's tendencies was relieved in a letter she sent him on February 8, 1882, RC-TLM.

Theo sent Gifford a letter from Coalburg on May 28, 1882, RC-TLM, thanking him for the Alpha Delta Phi wedding gift. The vase remains today in the possession of the Willis family at Coalburg. The description of Edith's wedding ceremony is taken from a letter Edith's mother Katherine T. Edwards sent to her sister Sarah S. Tappan on June 3, 1882 (WC). Cathy Raitt, great grandniece of Edith Edwards Mead, kindly provided this transcript.

CHAPTER 15: Honeymoon in England

Iceberg activity in the Atlantic was reported as "Icebergs in the Atlantic," and "Icebergs Dangerously Near," in *The New York Times*, June 2 and June 14, 1882. His father's letter hoping he had "escaped these annoyances" is dated June 26, 1882, RC-TLM.

Buxton was one of the earliest British spa towns where natural mineral water emerged from a group of springs at a constant temperature of 82 degrees

Fahrenheit. It attracted wealthy visitors who drank the medicinal waters and took part in the many hydrotherapy treatments.

TLM's letters to his parents, all in RC-TLM, help track their activities and movements: "went to the Zoo," July 6, 1882; "Chamber of Horrors and Kew," July 18, 1882; and "tired of the city," August 16, 1882. His mother's warning of the alcoholic temptations of London come from a letter she sent from New York, August 7, 1882, RC-TLM.

Details of Webster Smith and the photograph with Anne Edwards are from a personal communication with Cathy Raitt. "Coalburg scenery more beautiful than seen in England" is in TLM to parents, October 13, 1882, RC-TLM.

CHAPTER 16: The Eustis Years

Harry Norton completed the budding process in June 1883 with 228 buds of various types as documented in TLM to parents, June 3, 1883, RC-TLM. The fruit trees and pineapples had already been planted in August 1881, as recorded in father's diary entry, August 24, RC-TLM.

Family letters, RC-TLM, over the summer of 1883, have the Meads at Coalburg and the Catskills, and Theo at Ithaca in early October. They returned to Florida with Edith's mother who helped in clearing up and putting down matting—TLM to parents, November 1, 1883, RC-TLM. Mrs. Edwards' first impression of Florida is captured in TLM to parents, November 7, 1883, RC-TLM. With the plant losses after his return from the North, Theo had to rethink his approach to pineapple growing and decided on a protective pineapple pit as disclosed to his parents, December 23, 1883, RC-TLM.

Theo's initial success with citrus growing is contained in TLM to his parents, March 11, 1884, RC-TLM. His inflated judgment of an income of $40,000 comes from TLM to parents, March 8, 1884, RC-TLM.

Mrs. Edwards' departing comments to Theo as "an advertisement for Florida" is referred to in TLM to parents, February 7, 1884, RC-TLM.

Although the Civil War had been over for almost twenty years when Theo came to Florida, many Southerners, black and white, labeled Northerners seeking economic opportunity in the South as carpetbaggers, who were not to be trusted. It would be the early twentieth century before such resentment began to dissipate. Theo voiced these concerns over the difficulty of getting servant help in TLM to parents, June 17, 1884, RC-TLM.

Theo was a keen grower of exotic palms, which he sourced from all over the world (see Mead, *Autobiography*, 9). *Erythea edulis* is now *Brahea edulis*, the Guadalupe Palm; *Cocos flexuosa* is now *Syagrus flexuosa* and was one of Theo's favorite palms (TLM to parents, November 13, 1883, RC-TLM). *Cocos bonnetii* is renamed *Butia capitata*, the Pindo or Jelly Palm, whose fruit are sweet with a pineapple/banana flavor. Palm germination from seed could be very slow as he recounted to his parents, July 19, 1885, RC-TLM, with *Acrocomia totai* holding the record (Mead, *Autobiography*, 9).

Edith did not like the name *Crescent* in the farm address, considering it rather common (TLM to parents, February 4, 1884, RC-TLM), so Theo changed it to *Ponemah*, from the home of departed souls in Longfellow's *Song of Hiawatha*.

The description of Theo's garden and nursery in 1885 is gleaned from a raft of letters in the Rollins College Archives; the one on July 19 referring to his *Eucalyptus* collection. Theo's record keeping impressed his parents as stated in father to TLM, March 7, 1884, RC-TLM.

W. J. Holland was America's great popularizer of butterflies and moths in the first half of the twentieth century, and the author of two best-selling books at the time, *The Butterfly Book* (1898) and *The Moth Book* (1903). He supported active collectors worldwide and bought up many major collections from previously uncollected regions between 1890 and 1930. His visit with Theo at Eustis Heights is recorded in TLM to parents, March 6, 1884, RC-TLM. Eventually Holland donated his private collection of over 250,000 specimens to the Carnegie Museum of Natural History in Pittsburgh, wherein Mead's collection resides today.

A typical day at Eustis Bluff is assembled from a patchwork of items found in the letters TLM sent to his parents in 1883, 1884 and 1885 (RC-TLM, Box 10, folders 16, 17 & 18), and information about Eustis from Akright and McClellan, *Eustis*.

Early theories were that malaria was caused by bad air ('mala aria' in Italian). It was not until the end of the nineteenth century (1897) that the cause of the infection was proved the bite of infected mosquitoes.

Ice cream was a weekly treat as described in TLM to parents, July 7, 1885, RC-TLM. The gripping tale of Dr. Jekyll and Mr. Hyde comes from TLM to parents, January 27, 1885, RC-TLM. Theo's opinion of the majority of his Eustis neighbors is revealed in TLM to parents, August 23, 1885, RC-TLM.

Harry Robie wrote to Theo on February 22, 1883, RC-TLM, asking about citrus prospects in Central Florida. Robie's grove was on the north side of what is now Robie Avenue, to the east of present day US441/FL500.

CHAPTER 17: Dr. Henry Foster

The description of Henry Foster's belief in Christian principles and establishment of the sanitarium is from Samuel Hawley Adams, *Life of Henry Foster, Founder Clifton Springs Sanitarium* (London: Forgotten Books, 2013), 160–161.

The notice of Henry and Mary's marriage is contained in Ontario County New York, Marriage and Death Notices published in Ontario Repository & Messenger – Canandaigua, NY, 1872–1873, http://ontario.nygenweb.net/ontariorepositorymessenger2.htm, [accessed 28 January 2016].

Foster's arrival in the Lake Jesup community and his treatment of Walter Gwynn's wife appears in Richard Adicks and Donna Neely, *Oviedo – Biography of a Town* (New Smyrna Beach, FL: Luthers Publishing, 2007), 27.

Theo's first visit to the Gee Hammock and the excursion on Dr. Foster's steam yacht is described in a letter to Edith, April 1, 1882, RC-TLM. "Better than all the waters of Eustis," is from his second trip in TLM to parents, March 13, 1884, RC-TLM.

Theo's calculation of muck savings of $125 per acre is from TLM to parents, June 13, 1883, RC-TLM. The analysis of the annual income as a fraction of the running costs appears in TLM to parents, August 4, 1885, RC-TLM. His fruit and plant sales are declared in TLM to parents, September 4, 1885, RC-TLM—*Antholyza* is now renamed *Crocosmia*.

Theo's letter to *The Florida Dispatch* describing the effects of cold weather on palm survival appeared in January 25, 1886, volume 5, number 4; http://ufdc.ufl.edu/UF00055757/00005, [accessed 28 January 2016].

After an initial period of elation towards citrus growing, discouragement began to set in as Theo calculated the meager profit margin with oranges at $1.75 a box (TLM to parents, November 12, 1885, RC-TLM). He turned his gloominess against the region and inhabitants in a letter to his parents dated August 23, 1885, RC-TLM, and against the soil in one dated July 28, 1885, RC-TLM. His father's response suggesting selling the Eustis property came on August 19, 1885, RC-TLM.

His father's love of the Eustis Bluff area is contained in a letter to TLM, March 24, 1884, RC-TLM, and their desire as parents to be near him if he relocated in one dated November 17, 1885, RC-TLM.

CHAPTER 18: New Beginnings & Birth of a Daughter

The Gee Hammock description as one of the most beautiful groves in Florida comes from Oliver M. Crosby, *Florida Facts both Bright and Blue* (New York, 1887), 58. Theo was enthusiastic about the Lake Charm region (letter to parents, February 23, 1886, RC-TLM), and presented his analysis of the returns from Foster's 50 acres in a letter dated February 16, 1886, RC-TLM. Foster's deal with the Sanford and Indian River Railroad appears in Adicks and Neely, *Oviedo*, 30.

"Of absorbing interest," and "Do not care a copper," are from letters his father sent on February 27 and 26, 1886, respectively (RC-TLM). His father's diary entries for March 1886 are from RC-TLM. The Cater Grove purchase of $13,050 was officially recorded on April 28, 1886 as a 35 acre plot. Before this, on March 23,

Theo had purchased from the Fosters for $1,750 a roughly triangular adjacent front lot of 4.5 acres plus pond, close to the Lake Charm station. Both purchases reflected the current price for good citrus growing land of approximately $400 per acre. On October 28, 1886, Theo bought a further five-acre strip from Atkinson for $850. These property records are on the Orange County, Florida, Comptroller's website at http://or.occompt.com/recorder/eagleweb/docSearch.jsp., [accessed 28 January 2016].

His father's suggestion of buying the Eustis property off him for $16,000, and the sale of the Madison Avenue house for $42K cash are contained in letters sent on April 7 and 8, 1886, RC-TLM.

Dr. Langdon's recommendation of a sea voyage is recounted in a letter from Edith to TLM, June 24, 1886, RC-TLM. Theo told his parents how much stronger Edith was following her treatment in a letter, July 6, 1886, RC-TLM, and of "sloughing off my Eustis skin" in TLM to parents, August 24, 1886. His exuberance of his new life at Lake Charm, which extended to the "nicest way of earning a living" quote, is from TLM to parents, December 15, 1886, RC-TLM.

Theo was keen to buy more citrus grove land from Jelks, a neighbor, and made the case for purchase to his parents in a letter dated November 29, 1886, RC-TLM. His prediction of a minimum income of $10,000 a year is financial nonsense; the most they ever made in one year was $7,000 and the 15-year average of citrus returns from their groves was $2,500. Nevertheless, his mother decided to purchase the grove for him.

His father's continued love of the Eustis Bluff location is referred to in a letter dated March 20, 1887, RC-TLM.

The scarcity of mandarins that year is disclosed in TLM to parents, December 13, 1886, RC-TLM. The Florida Fruit Exchange returns in March 1887 come from letter TLM to parents, March 23, 1887, RC-TLM.

No primary Mead source has been found explaining the origin of the name *Waitabit*, but according to a 1934 unpublished article by Helen Golloway,

(*Theodore L. Mead*, typed manuscript, RC-TLM, Box 1, folder 4), the name came about because with limited funds they first decided to build a small home and planned on building a larger one later. A detailed description of the decorations in the parlor is from TLM to parents, June 22, 1887, RC-TLM.

The decision to go to Montclair to be near Julia Inness appears in TLM to parents, April 25, 1887, RC-TLM. Julia Goodrich Smith was the daughter of wealthy publisher Roswell Smith, who owned the prominent New York publishing company *The Century Company*. She married George Inness, Jr (1854–1926), one of America's foremost figure and landscape artists and the son of George Inness Sr., himself a distinguished American landscape painter. In the 1880s, they lived in Roswell Manor, Montclair, and in the early twentieth century built a winter home in Tarpon Springs, Florida, where they were prominent residents.

The circumstances surrounding Dorothy's birth come from TLM to parents, October 4, 1887, RC-TLM. *Hinc illae lacrimae* in Latin means "hence these tears." His parent's movements to see the baby and attend the baptism were recorded in his father's 1887 diary of October 5 and November 7, RC-TLM.

Dorothy's early feeding problems are disclosed in Edith to TLM, November 12, 1887, RC-TLM. Edith's sister Anne had given birth to her second child, Emily, on August 23, 1887, RC-TLM, so could help with the nursing. "Nothing of the Edwards except the ears," is found in Edith to TLM, November 14, 1887, RC-TLM.

Theo's advertisement for the sale of Royal Palms appeared in several issues of *The Florida Dispatch*—see for example Vol. 7, no. 49, December 5, 1887, 1,020. http://ufdc.ufl.edu/UF00055757/00102, [accessed 28 January 2016].

CHAPTER 19: Improving Lake Charm & Bringing up Dorothy

The prediction of a good citrus harvest comes from the Oviedo *Chronicle*, August 14, 1887. The details of the 1887 citrus returns from Theo's groves come from a later letter (TLM to parents, June 5, 1894, RC-TLM), summarizing the returns and fertilizer costs in graphical form for the years 1887 to 1894.

The narrative of the Oviedo, Lake Charm & Lake Jesup Railroad is assembled from Adams, *Henry Foster*, 70; Adicks and Neely, *Oviedo*, 30–32; and the papers in Box 5, folders 1–9, RC-TLM. Additional information comes from Jim Robinson, "Lack of Rail Nearly Uprooted County Seat", June 23, 2002, *Orlando Sentinel*. http://articles.orlandosentinel.com/2002-06-23/news/0206210515_1_oviedo-lake-jesup-orlando, [accessed 15 January 2016]. The sketch showing the proposed railroad route around Theo's property comes from TLM to parents, February 1888, RC-TLM.

The Lake Charm Improvement Company story comes from Adams, *Henry Foster*, 71; Adicks and Neely, *Oviedo*, 29; and the papers in Box 5, folders 10–13, RC-TLM.

Theo's letters in February 1888 to his parents (RC-TLM), tell of Dorothy's weight gain and appearance with "bright eyes and rosy cheeks." His report on her condition on returning from St. Augustine is from TLM to parents, June 26, 1888, RC-TLM.

In the late nineteenth century, diseases such as yellow fever, colloquially known as 'yellow jack' after the yellow flags ships flew if they were infected, could spread rapidly, decimating populations of cities and causing panic among the citizens. On July 28, 1888, Richard McCormick, "a saloonkeeper and otherwise disreputable person," brought yellow fever to Jacksonville by riding a train from Tampa; source, R. Scott Huffard, Jr., "Infected rails: yellow fever and southern railroads," *Journal of Southern History* 79.1 (February 2013), 80. Before the disease ran its course, nearly 5,000 of the 14,000 people remaining in Jacksonville caught it and more than 400 died.

The description of running the yellow fever quarantine gauntlet around Jacksonville comes from Edith to TLM, October 28, 1888, RC-TLM. At Waycross, Georgia, mail fumigation was carried out in a converted boxcar, using sulfuric fumes cooked up from the large iron kettle inside the car as disinfectant. Source: Herbert P. McNeal, *Yellow Fever Mail from the 1888 Florida Epidemic*, Florida Postal History, Journal Vol 1, No. 1. July 1993, 27–32.

Edith to TLM, October 31, 1888, RC-TLM, speaks of Dorothy's excitable state, and on November 19, 1888, RC-TLM, her continued unmanageable behavior.

Overhearing Catherine praying for Theo not to catch yellow fever comes from Edith to TLM, November 16, 1888, RC-TLM.

CHAPTER 20: Bumper Citrus Harvests

Edith's delight at receiving the photographs of Dorothy and her longing to go back to *Waitabit* come from Edith to TLM, February 9, 1890, RC-TLM.

Theo's citrus returns are from TLM to parents, June 5, 1894, RC-TLM. Individual tree yields and the prodigious yield of one of them in 1886 appears in TLM to parents, December 9 and 17, 1886, RC-TLM. The economics of citrus picking and the grading system used are described in TLM to parents, December 13 and 15, 1886, RC-TLM. The "Friend of the russet" comment appears in the Oviedo *Chronicle*, February 19, 1889.

Theo's decision to make his own citrus fertilizer is contained in TLM to parents, November 25, 1888, RC-TLM. Theo's order for a 'Murray' buggy was placed on September 26, 1890, RC-TLM.

The various difficulties with Dorothy are described in TLM to parents, December 12, 1890, and February 19, 1891, RC-TLM. The plan to go north over the summer that had Theo "glad to pack both of them off," is from TLM to parents, April 19, 1891, RC-TLM.

Edith's news from Roswell Manor comes from a letter dated July 9, 1891, RC-TLM. Dorothy's excitable state continued, according to Edith to TLM, September 2, 1891, RC-TLM.

TLM to parents, July 12, 1891, RC-TLM, contains an account of his first recorded orchid hybridization experiments.

Theo's reaction to meeting up with Edith and Dorothy at Coalburg are reported in TLM to parents, October 9, 1891, RC-TLM. Anne's hunt for a baby to adopt is made known in TLM to parents, October 8, 1891, RC-TLM. News of the adoption of Harold appears in letters to his parents dated November 20 and 22, 1891, RC-TLM.

Back at Lake Charm, the tiredness and problems with Harold that made them both miserable are revealed in TLM to parents, December 10 and 11, 1891, RC-TLM.

CHAPTER 21: Scarlet Fever Strikes

Theo informed his father at Eustis Bluff that Dorothy was ill on February 10, and that the diagnosis was scarlet fever on the 12th (father's diary, 1892, RC-TLM). Scarlet fever, characterized by a reddened sore throat (known today colloquially as strep throat), red cheeks and a rash, was one of the most common nineteenth century childhood infectious diseases in the United States, causing fatalities of 30% in some areas.

Theo kept his parents well-informed of the progress of Dorothy's illness with letters (all RC-TLM). The relevant ones here were sent on February 12 ("milk is so good, Mamma!"); February 13 (Harold boards with Christiansens); February 15 ("we are as anxious as possible"); February 17 ("opened her eyes a few times"); February 18 ("as well as might be expected"); February 23 ("barely registered it was his birthday"); February 24 ("like a frightened wild animal"); and February 25 (telegram sent).

The receipt of the telegram and details of Dorothy's burial were recorded in his father's diary of February 24–28, 1892, RC-TLM.

The Reverend Lyman Phelps was an Episcopalian minister who lived at Sanford and was a keen horticulturist and member of the Florida State Horticultural Society, visiting Theo at Lake Charm on a number of occasions.

The planting on a tree azalea at Dorothy's grave comes from Adicks and Neely, *Oviedo*, 49. Theo's mother and father's reaction to the tragedy are from a letter dated March 3, 1892, RC-TLM.

The cocoa butter and clock stories about Dorothy are disclosed in TLM to parents, June 11, 1892, RC-TLM. Theo's reaction in a church service to the loss of Dorothy is from TLM to parents, March 27, 1892, RC-TLM.

The result of Edith's examination by Dr. Foster is contained in TLM to parents, September 17, 1892, RC-TLM. The quotation the "only thing (to) make life worth living," comes from TLM to parents, March 6, 1892, RC-TLM. His father's response is captured in a letter dated September 1, 1892, RC-TLM.

The Robie's decision to name their second child after Theo comes from TLM to parents, June 8, 1892, RC-TLM.

Generally, Theo's parents spent part of their summers at Chautauqua gatherings, a format that suited them perfectly as supporters of adult education and those with a love of learning.

The decision to spend the summer in North Carolina and the fate of Harold is told in TLM to parents, June 15 and August 10, 1892, RC-TLM. The act of handing Harold back to the home in Charleston appears in TLM to parents, September 15, 1892, RC-TLM.

"Can't make things for the remainder of the human race," is taken from TLM to parents, December 20, 1892, RC-TLM; "Theodore Mead Robie" in TLM to parents, December 29, 1892, RC-TLM; and "Happiest year of her life" in TLM to parents, January 3, 1893, RC-TLM.

CHAPTER 22: Irrigation & The Great Freeze

The plan to install a steam-pumped irrigation system is described in TLM to parents, September 6, 1891, March 6, 1892, and March 13, 1892 (all RC-TLM). The technical details and performance of his irrigation plant were written up in *The Florida Dispatch, Farmer & Fruit-Grower*, July 7, 1892, 530, and in T. L. Mead, "A Frost-Proof Orange Orchard," *Country Life in America* 7 (February 1905): 367–369 & 385–386.

Theo's contract to irrigate his neighbor's groves comes from TLM to parents, April 11, 1893, RC-TLM.

Edith's headache episode is divulged in TLM to parents, January 22, 1893, RC-TLM. The idea that living in Florida affected female fertility was an unusual viewpoint from someone who was normally logical and relied on scientific reason, but is postulated in TLM to parents, May 13, 1893, RC-TLM.

Edith to TLM, June 29, 1893, RC-TLM, contains her mother's puzzlement over the need for Clifton Springs treatment. Dr. Foster's examination results were communicated in Edith to TLM, July 10, 1893, RC-TLM. Edith's lament concerning the sacrifice of being apart comes from Edith to TLM, July 16, 1893, RC-TLM.

Letters "written in the Lord's time" always brought a rebuke from Theo's mother—letter to TLM, June 28, 1893, RC-TLM. Theo's side-trip to Ithaca is taken from TLM to parents, November 1, 1893, RC-TLM. The situation back at the house in Lake Charm comes from TLM to parents, November 5, 1893, RC-TLM. The account of the flowering of the night-blooming cereus is given in TLM to parents, May 15, 1894, RC-TLM.

His father's fear "of being busted" is in a letter to TLM, November 28, 1893, RC-TLM. Edith's piano-playing abilities come from TLM to parents, May 30, 1894, RC-TLM. From Coalburg, Edith wrote to Theo's mother expressing her delight in the snowy weather—letter dated December 26, 1894, RC-TLM.

The description of the first December freeze is based on John A. Attaway, *A History of Florida Citrus Freezes* (Lake Alfred, FL: Florida Science Source, 1997), 48–49, where there is some material taken from George Rainsford Fairbanks, "Florida: Its History & Romance" (Jacksonville: H. and W. B. Drew Co., 1898), 232.

The reference to a $2M loss appears in Jeff Cannon, "The Great Freeze of 1894–1895," New Port Richey Patch, http://patch.com/florida/newportrichey/the-great-freeze-of-1894-1895, [accessed 15 January 2016]. Theo's description of the effects of the freeze at Lake Charm and the possibility of substantial returns in the future is contained in TLM to parents, January 16, 1895, RC-TLM.

The February freeze temperatures and the simile of citrus trees making sounds like the cracking of walnuts are from Attaway, *History of Citrus Freezes*, 51 & 54. The effects of the freeze on Theo's groves are described in TLM to parents, February 7, 1895, RC-TLM. "I stood upon the bank" comes from J. A. Hendley, "History of Pasco County Florida," page 34 in subarticle "Colonel Hendley Recites His Record," http://digital.lib.usf.edu/content/SF/S0/03/64/28/00001/C54-00012.pdf, [accessed 20 December 2015]. The hasty abandonment of many settlers is told in Attaway, *History of Citrus Freezes*, 51.

The 97% drop in citrus shipments comes from Benjamin Reilly, *Tropical Surge: A History of Ambition and Disaster on the Florida Shore* (Sarasota: Pineapple Press, 2005), 85. Income falling from $100 a week to a handful of dollars appears in Mead, *Autobiography*, 10. News of the Markham Grove sale is from TLM to parents, June 12, 1895, RC-TLM. The Markham property sold for $4,000 at a sheriff's sale according to TLM to parents, November 7, 1895, RC-TLM. The tragedy of Mr. Markham is recounted in Attaway, *History of Citrus Freezes*, 53.

CHAPTER 23: Humanitarian Efforts in the Community

The herbal remedy Drosera was effective in the treatment of a number of health conditions, such as violent coughs, particularly bronchitis and whooping cough, and behavioral disorders, especially among children. "Edith in great demand" is from TLM to parents, July 21, 1895, RC-TLM, while treating Mrs. Brock's whooping cough is mentioned in TLM to parents, July 30, 1895, RC-TLM.

Some historical references, but none of the primary sources, refer to Theo as "Dr. Mead." There is no record of him receiving this as an honorary title or earning it as a doctorate degree from any educational establishment. It is likely that this was an informal title given to him by the many Oviedo families that the Meads treated.

The account of how busy Edith was making presents for Christmas is from TLM to parents, December 9, 1895, RC-TLM. The presents they received on Christmas Day are from a single sheet manuscript in RC-TLM.

In Florida, and in many of the Southern States, it was usual in the latter parts of the nineteenth century for farmers to let their hogs range freely over the countryside to find their own forage—see Ian Frazier, "Hogs Wild" *The New Yorker*, 71–72 (December 12, 2005). Theo's creative solution of poisoning is divulged in TLM to parents, July 6, 1894, RC-TLM. Despite this epicurean diet, the hogs managed to destroy his okra plants and caladium tubers—TLM to parents, November 21 and December 22, 1895, RC-TLM.

Edith's predicament in having to choose between caretaking her family and supporting her husband is apparent in Edith to TLM, October 14, 1896, RC-TLM.

The craze of bicycling introduced thousands of people to individual and independent mechanical transportation before the automobile, and the impact on female emancipation was considerable. The safety bicycle gave women mobility and helped trigger a revolution in dress in the form of the then-shocking bloomers; hence her response, Edith to TLM, October 11, 1896, RC-TLM. "Wheel" was a nineteenth-century American term for a bicycle.

Archie Foss was Theo's nephew, being the son of his aunt, Caroline H. Mead, who married the Reverend Archibald C. Foss. The Eagle Bicycle Manufacturing Company based in Torrington, Connecticut, built bicycles from 1888 to 1900. "Dodging through the bushes" is a quote from TLM to parents, May 17, 1897, RC-TLM.

Information on decorating the chapel on Sunday comes from TLM diary entries for 1904 and 1910 (WC), and TLM to parents, March 24, 1899, RC-TLM.

CHAPTER 24: Financial Problems

Theo's admission that he felt strongly about providing income for them to live on is revealed in TLM to parents, March 11, 1898, RC-TLM. Sales of his standard caladium tubers only brought in a moderate amount (TLM to parents, May 3, 1897, RC-TLM), and the real money came from rare and unusual types (TLM to parents, November 2, 1895, RC-TLM).

Repayment of the loan is from TLM to parents, December 21, 1897, RC-TLM.

The winter of 1899 as "a little bit too ridiculous," appears in TLM to parents, February 10, 1899, RC-TLM.

Number 14, W. Fiftieth Street, at that time rented, was part of the Luqueer family property portfolio. J. W. Overstreet was a founder member of the Overstreet Turpentine Company, one of many turpentine producers in Central Florida.

The $35,000 represented Lake Charm income from 27,726 boxes of citrus for 10½ years, from 1886 to 1897, averaging $2,400 per annum—TLM to parents, unrecorded date, presumed 1898, RC-TLM. Theo's fears of an impending financial calamity are from TLM to parents, April 7, 1899, RC-TLM.

"Extinct as the mastodon" is from TLM to parents, April 14, 1899, RC-TLM; "shirtsleeves to shirtsleeves" from TLM to parents, February 10, 1899, RC-TLM.

The Robie's intention to quit Florida is from TLM to parents, March 22, 1899, RC-TLM. The Mead's decision to see them before they left comes from TLM to parents, March 27, 1899, RC-TLM. Theo confessed his closeness to the Robie boys in TLM to parents, March 24, 1899, RC-TLM.

His father's love of the Eustis Bluff locality appears in letter to TLM, November 14, 1897, RC-TLM. The description of the circumstances of his father's death at the time is contained in a letter from TLM to the Florida State Health Officer, January 21, 1901, RC-TLM. His father's remains were not disinterred from Eustis Bluff until ten years later and moved to Green-Wood cemetery in

Brooklyn—TLM diary entry, August 6, 1901, WC. Theo learned the true reason for his death many years later.

Mary Mead's living arrangements for 1910 come from United States Federal Census, New York, State Census, Kings, Brooklyn, 4th Assembly District, 21st Election District, Block B, page 10.

Will Edwards gave his full support to his brother-in-law "as long as he lived"—letter Will to TLM, January 20, 1900, RC-TLM. His proposal for a royalty investment in his businesses comes from Will to TLM, October 27, 1902, RC-TLM. The suggested investments in the West Virginia coal and gas fields elicited Theo's grateful thanks in TLM to Will, March 19, 1900, RC-TLM.

The letter inviting Theo to an all-expense paid grand reunion of Alpha Delta Phi fraternity in New York is McVoy to TLM, May 6, 1901, RC-TLM. At this stage, Theo was a well-loved and venerable member of the fraternity who had a strong melodic voice and enjoyed singing. His party piece was a rendition of "Taranty my son", a traditional New England ballad about a boy poisoned by his grandmother by being fed snakes rather than eels for supper.

The Lake of the Woods 800-acre plot, previously leased for 3.5 years in 1897 to Overstreet for turpentine production, was disposed of on October 2, 1901.

CHAPTER 25: Acceptance of Faith

Theo's experiences in the Episcopal Church are from TLM to parents, April 18, 1895, RC-TLM. The appropriate sermon appears in Phillips Brooks, *The Light of the World and Other Sermons* (New York: E. P. Dutton & Company, 1910), 52. Romanes' conclusion that it was "reasonable to be a Christian believer" is from George John Romanes, *Thoughts on Religion* (Chicago: The Open Court Publishing Company, 1895), 196. W. H. Savage's *The Intellectual Basis of Faith* is an address in M. J. Savage, *Belief in God* (Boston: G. H. Ellis, 1881), 153–176.

Theo tried to explain his thought processes in TLM to parents, May 20, 1895, RC-TLM. His objection to the religious 'war dance' of his mother is in TLM to parents, June 5, 1895, RC-TLM. "Not making a fetish out of the Bible" comes from TLM to parents, May 29, 1895, RC-TLM, as does the "gaping peasants of Judea" quote.

His position regarding dogmas is extracted from TLM to parents, May 22, 1895, RC-TLM, and his attraction to the Episcopal faith in TLM to parents, June 5, 1895, RC-TLM. "Punishment from the Lord" was frequently used by his mother in their religious exchanges, this from an earlier one, Mary Mead to TLM, October 4, 1892, RC-TLM. Theo's plead for a compromise is found in TLM to parents, June 6, 1895, RC-TLM.

We do not know the effect on Theo of this extract from a letter from his mother, but one of Theo's character weaknesses was his intense sensitivity to criticism, particularly from those he loved. His mother was not content with Theo having found God's love in his own way, and so telling him that his life's horticultural work was still a minor achievement and no cause for rejoicing must have hurt deeply. Theo cut this part of the letter out and returned it with one of the sayings of St. Paul, in TLM to parents, April 4, 1896, RC-TLM.

Evangelical religion was hard-wired into his mother's DNA, and nothing that Theo said in this quotation (TLM to parents, May 31, 1896, RC-TLM) changed his mother's view, and vice versa.

In October 1892, the Missionary Jurisdiction of South Florida was created with William Crane Gray as its first bishop. Bishop Gray wrote to TLM on April 16, 1903, RC. Details of the service are from "Missionary Jurisdiction of South Florida", page 59, in the archives of The Cathedral Church of St Luke, Orlando. His mother's congratulations are in Mary Mead to TLM, May 23, 1903, RC-TLM.

As Theo explained in his autobiography, the confirmation rated as one of the three wonder days of his life, together with his initiation into the Alpha Delta Phi fraternity and his marriage to Edith. The letter extracts "exactly as Phillips Brooks has written," are from TLM to Mott Williams, May 23, 1903, RC-TLM; and "every right to trust in his loving kindness" from TLM to Julia Inness, June 14, 1903, RC-TLM.

CHAPTER 26: Medical Issues & Loss of Family Members

The result of the doctor's examination of Edith's hand is from Edith to TLM, October 7, 1902, RC-TLM. Will Edwards married Miss Hope M. Christensen, daughter of Christian T. Christensen of Brooklyn, New York, in London on July 5, 1902.

Theo's mother was fond of invoking the wrath of the Almighty to explain why unpleasant things happened to Theo, such as when he caught a severe cold—mother to TLM, May 27, 1907, RC-TLM.

The migration of cultured Northerners from Lake Charm since the Great Freeze had been considerable, as Theo remarked in TLM to mother, February 13, 1900, RC-TLM.

Edith's report on the state of health of her father is from Edith to TLM, August 6, 1907, RC-TLM. Edith's treatment for her hand at Clifton Springs is described in Edith to TLM, August 30, 1907, RC-TLM. Fretting about Theo's mother comes from Edith to TLM, September 3, 1907, RC-TLM.

The "Infernal Machine" poster is from the Mead papers at Rollins College. Despite the warning, the home still ended up being broken into periodically (Fields to TLM, February 16, 1909, RC-TLM).

The narrative of Mary Mead studying Greek appeared as "Woman a Greek Student when 85 years of Age," *New York Herald*, May 26, 1909. The state of Mr. W. H. Edwards' study and financial affairs following his death were taken from Will Edwards to Edith, July 4, 1909, WC. The talents of Edith's sister Anne in painting the Cherokee rose hedge come from TLM to parents, April 16, 1893, RC-TLM.

The organization of Edith's trip to Italy with her sister Anne is from Will Edwards to Edith, August 31, 1910, WC, and Will Edwards to TLM, November 1, 1910, RC-TLM.

The Battle Creek Sanitarium, founded by John Harvey Kellogg, was run on holistic grounds focusing on nutrition, enemas, and exercise. Mary Mead's letter to Theo saying she was unwell is dated January 21, 1914, RC-TLM. The Kings County Surrogate's Court document records Theodore L. Mead as a petitioner on November 20, 1914, RC-TLM.

The insight into the musical tastes of Theo and Edith, and the Victrola machine, appear in TLM to Clarence Gilbert (one of Theo's penpals), undated, RC-TLM.

The thought that financial pressures had hastened the death of Will Edwards comes from Julius Seymour to TLM, January 20, 1916, RC-TLM.

Data concerned the Oviedo Woman's Club comes from Adicks and Neely, *Oviedo*, 68; http://www.oviedowomansclub.org/our-community/our-history.html, [accessed 28 January 2016]; and Jack Tilton to TLM, March 26, 1916, RC-TLM.

The description of the severe nose and ear infection is from TLM to Robert Rolfe, November 23, 1916, RC-TLM.

CHAPTER 27: Orchid Breeding

The earliest record of a Mead orchid cross appears to be in TLM to parents, July 12, 1891, RC-TLM. The "unpromising bundles of sticks" quote is from "Raising Orchids in Florida," a written presentation by Mead given in 1923 (manuscript in Rollins College Archives). Theo's crosses during 1893 and 1894 are documented in a letter to H. Clinkaberry, April 12, 1895, RC-TLM.

In 1895, orchid seed germination was not well understood; it would be the end of the century before Bernard finally established that it required mycorrhizal fungus as a critical symbiotic element (N. Bernard "Sur la germination du Neottia nidus-avis," *Comptes Rendu Hebdomadaire des Séances de l'Académie des Sciences*, Paris, Vol. 128: (1899): 1253–1255). Theo's attempts at improving the rate of germination are described in a number of letters to his parents in the 1894 to 1897 period (RC-TLM).

One of Mead's main notebooks recording his orchid and bromeliad crosses from 1904 to 1936 is part of the MAS-TLM collection. The Helen G. Connery

Collection consists of 115 of Mead's glass plate negatives, the majority of which are of orchids (WPPL). An album containing some of the hand-colored prints of orchids and other subjects resides in RC-TLM, Box 3.

Mead's orchid, *Cattleya* Ludbrosa, was finally recognised in 2014. The official record appears as an update to *The Orchid Review* (Quarterly Supplement to the International Register and Checklist of Orchid Hybrids (Sander's List), January–March 2014 Registrations, distributed with Volume 122, Number 1306, June 2014).

Edward Owen Orpet (1863–1956) was an English plantsman who came to America and for more than twenty years was superintendent of the estate of E. V. R. Thayer in South Lancaster, Massachusetts. In October 1900, he exhibited a collection of hybrid *Cattleya*, which he had raised from seed, at the Massachusetts Horticultural Show. In 1922, he was appointed Superintendent of Parks for Santa Barbara, California. Today much of the horticultural beauty of this city is due to his talents and his legacy lives on in the city's Orpet Park.

Oakes Ames (1874–1950) was an American botanist specializing in orchids. He built up an extensive orchid herbarium, which he left to Harvard University. His estate is now the Borderland State Park in Massachusetts.

Robert Allen Rolfe (1855–1921) started at the Royal Botanic Gardens, Kew in London as a gardener in 1879 and became a knowledgeable and experienced horticulturist with a particular interest in orchids. He founded the world's first specific orchid publication, *The Orchid Review*, and edited it for twenty-eight years.

The key reference book for orchid hybridization at that time was Robert A. Rolfe and Charles C. Hurst, *The Orchid Stud Book* (London: Frank Leslie, 1909). The list of hybrids unique to Theo come from a single page undated manuscript, RC-TLM.

The extent of the cooperative work between Lewis Knudson and Theodore Mead is captured in many letters in RC-TLM. That Knudson's work would likely not have succeeded without Mead's inputs, comes from T. W. Yam and others, "Orchid seeds and their germination: A historical account," in *Orchid Biology: Reviews and Perspectives*, ed. T. Kull and J. Arditti, 483 (Dordrecht: Kluwer, 2002).

Knudson's paper appeared as Lewis Knudson, "Nonsymbiotic Germination of Orchid Seeds," *Botanical Gazette* 73, no. 1 (1922): 1–25. The established protocol in research papers is to name co-workers as co-authors when they have made a significant and intellectual contribution to a particular work, as Mead did.

Nonsymbiotic germination provided the keys to developing the orchid industry as we know it today—R. J. Griesbach, "Development of *Phalaenopsis* Orchids for the Mass-Market," in *Trends in New Crops and New Uses*, ed. J. Janick and A. Whipkey, 459 (Alexandria, VA: ASHS Press, 2002).

CHAPTER 28: Citrus Freeze Protection

"A cold wave from the Dakotas" is from W. S. Hart, "Grove Protection at Time of Cold Waves," in *Proceedings of the Eighteenth Annual Meeting of the Florida State Horticultural Society* 18 (May 1905): 109–114.

A description of Theo's thinking concerning cold weather protection of citrus is found in Mead, *Country Life in America*, 367. Julia Inness' help with the shed expenses is from TLM to mother, February 8, 1900, RC-TLM.

Theo's citrus tree order is recorded in TLM to Conner Fruit Tree Provider, April 2, 1900, RC-TLM. His expectation of substantial returns appears in TLM to R. B. Williams, Alpha Delta Phi graduate correspondent, May 9, 1900, RC-TLM.

With the temperature falling at roughly one degree per hour, the alarm system gave Theo time to start up the pumps, as described in Mead, *Country Life in America*, 369.

The shed came into its own in the winter of 1904/5 as reported with before and after photographs in T. L. Mead, "Defying the Blizzard in a Florida Orange Shed," *Country Life in America* 8 (May 1905): 118 & 120.

Irrigation and other freeze protection methods are discussed in Dorota Z. Haman, ed. *Frost and Freeze Protection Workshop* (University of Florida, Florida Cooperative Extension Service, Fact Sheet HS-76, 1995). Overhead irrigation is a standard technique today for cold protection—see for example L. R. Parsons and B. J. Boman, "Microsprinkler Irrigation for Cold Protection of Florida Citrus," (University of Florida, ISAS Extension, Fact Sheet HS-931, 2003).

The amount collected from Theo's Christmas planting of cucumbers is from Adicks and Neely, *Oviedo*, 50. The eggplant numbers are listed on a single page manuscript dated 1906, RC-TLM.

Cyrus Miner Berry (1870–1941) was a fellow entomologist and member of The Florida Entomological Society of Gainesville. Through a partnership involving pooling their respective fruit, flower and vegetable growing experiences, and those of shed protection for tender plants, the Berrys and the Meads became firm friends.

Berry's experience in cloth greenhouse protection had been recorded in R. M. Fletcher Berry, "The Protection of Fruits and Vegetables in Florida – The Principles and Opportunities of Intensive Farming as Embodied in the Cloth Greenhouse System," *Country Life in America* 11 (January 1907): 315–316. Earlier in 1902, he was listed as co-inventor of a shed system for the cold weather protection of tender crops; Israel C. Putnam and Cyrus M. Berry, "Fruit or Vegetable Protector", US Patent US 711225 A, Granted 1902.

The virtues of diversification in vegetable growing and the data for the profitable year 1911 come from Berry to TLM, April 23, 1911, RC-TLM; the perils of concentrating on lettuce and the data for the profitable year 1915 is from TLM to E. N. Reasoner, January 6, 1915, RC-TLM.

CHAPTER 29: Caladium Growing with Henry Nehrling

The self-introduction based on the Mead article on *Pancratrium ovatum* in *The Florida Dispatch* is from Nehrling to TLM, October 30, 1890, RC-HN.

Nehrling asks detailed questions about Theo's plantings in Nehrling to TLM, December 14, 1893, RC-HN. An early interest for Nehrling was ornithology and he published two volumes of his book on birds, Volume 1 in 1893 and Volume 2 in 1896. Both Theodore Mead and Henry Nehrling were members of the Florida Audubon Society, and in 1912, Nehrling was appointed to the executive committee. The description of Theo's garden is from Henry Nehrling, *Our Native Birds of Song and Beauty* (Milwaukee: George Brumder, 1893), 1:43.

The letter Nehrling to TLM, April 28, 1903, RC-HN, tells of his settling in at Gotha. According to his diary entry, he first arrived at Gotha in January 1902 (Nehrling, Henry, "Diary No. 1, December 30, 1901, to September 11, 1904," Nehrling Papers, Special Collections and University Archives, University of Central Florida, Orlando, Florida).

The extensive hybridization of the caladium, much of it performed in the nineteenth century, is covered in Arno Nehrling, "The Fancy Leaved Caladiums" *The American Florist* 32 (July 1909): 1213–1215.

The recognition of Mead as the first hybridizer of the caladium in North America comes from Robert W. Read, ed., *Nehrling's Early Florida Gardens* (Gainesville, FL: University Press of Florida, 2001), 72. Nehrling's decision to specialize in fancy leaved caladiums and order named cultivars from a German grower is contained in Nehrling to TLM, January 2, 1905, RC-HN. Nehrling's extensive planting program is disclosed in "Caladiums - from the tropical rainforests to today's gardens," www.classiccaladiums.com/rainforest_plants_the_caladium_history.htm, [accessed 10 December 2015].

The ups and downs of caladium growing at Gotha come from Nehrling to TLM, November 8, 1907, RC-HN, and Robert W. Read, ed., *Nehrling's Plants, People and Places in Early Florida* (Gainesville, FL: University Press of Florida, 2001), 6.

It was common practice in nursery catalogs at the time not to acknowledge the hybridizer of new varieties. The lack of hybridizer knowledge is lamented in Arno Nehrling, *The American Florist*, 1215.

Hybrid caladiums attributed to Mead are taken from H. Nehrling "Bulbous and Tuberous Plants – Variegated Caladium", *Die Gartenwelt* 16 (January 1912): 4–5. A list showing those attributed to Nehrling comes from "Once Popular, Long Ignored – The Caladium," http://www.zone10.com/once-popular-long-ignored-the-caladium.html, [accessed 10 January 2016].

The letter suggesting the naming of Mead's new caladium *Mrs. Theodore L. Mead* is from Nehrling to TLM, June 22, 1906, RC-HN. The fate of the *T. L. Mead* hybrid was to be rebranded as *Blaze*. Nehrling and Mead were the founding fathers of the caladium industry in Florida worth an estimated $13M each year to the economy.

Theo's concern about the plethora of caladium hybrids is from TLM to Reasoner, January 6, 1915, RC-TLM. His starting parents for the arrow and lance-shaped caladiums are acknowledged in Mead, *Autobiography*, 13, and the text and photograph is from page 14, reference 81 "Arrow and Lance Caladiums, photographed August 2, 1924," in Box 3, RC-TLM. The tougher constitution of this new race of caladiums is commented on in Scott Ogden, *Garden Bulbs for the South* (Lanham, MD: Taylor Publishing, 1994), 206. Today they are commonly called 'strap' or sometimes 'lance' caladium.

Ten named arrow and lance caladiums hybridized by Mead are listed in Arno Nehrling, *The American Florist*, 1239. *Istachatta* comes from Nehrling to TLM, December 29, 1919, RC-HN.

The selling of the remainder of his caladiums is from W. Atlee Burpee to TLM, March 20, 1920, RC-TLM.

CHAPTER 30: Amaryllis

The first record of the hybridizing of the amaryllis is in TLM to parents, April 14, 1889, RC-TLM.

The extent of Nehrling's amaryllis collection is stated in Nehrling to TLM, November 8, 1907, RC-HN. The gift of a single bulb and access to pollen is from Wyndham Hayward, "Bulbous Plants Adapted to Florida," *Proceedings of the Florida State Horticultural Society* 61 (1948): 207.

The process with the red and white flowering amaryllis is described in Cecil Eastman, "Southern Personalities: Theodore L. Mead—Grower of Amaryllis and Orchids," *Holland's: The Magazine of the South*, July 1931, 7, 42–43.

The mystery of the Mead-strain amaryllis parentage was puzzled over in Wyndham Hayward, "The Mead Strain of the Nehrling Hybrid Amaryllis," *Year Book of American Amaryllis Society* 1 (1934): 62–63. Nehrling's orders for the Mead hybrid are from Nehrling to TLM, November 4, 1919, and January 28, 1920, RC-HN.

"Anxious to get back some of the wages," and the cut-price offer of bulbs at 10 cents each comes from TLM to E. N. Reasoner, January 6, 1915, RC-TLM. The City of Jacksonville order for 300 bulbs is from Elmo Acosta, Parks Commissioner to TLM, November 16, 1921, RC-TLM. The Vaughan Company catalog entry is from "Vaughan's Spring Flowering Bulbs 1924," page 8, http://www.biodiversitylibrary.org/item/155722#page/3/mode/1up, [accessed 28 January 2016].

The contract for supplying 3,000 amaryllis bulbs from Fall 1920 to Spring 1922 is from Vaughan Seed Store to TLM, April 10, 1920, RC-TLM. Moving amaryllis production to Sanford is mentioned in Mead, *Florida Trucker*, 4, and in TLM to E. T. Barnes, 1926, RC-TLM, when he talks about sowing 50,000 hand-crossed seeds.

Theo's amaryllis garden as the finest outside Government gardens is from Helen Golloway, "Theodore L. Mead", typed undated manuscript, Box 1, folder 4, RC-TLM. The fact that many of his garden visitors came long distances is from TLM to Ogden Willis, March 22, 1927, WC. The idea of putting up signs, and callers needing appointments, were suggestions from Madison Cooper to TLM, September 16, 1922, RC-TLM. Theo's admission that his generosity led to visitors choosing their own seedling is in Mead, *Autobiography* 11.

Notes of Grover's visits to the garden appear in TLM's diary of December 12, 1926, and August 28, 1927, WC. "I want to meet the famous Mr. Mead" comes from Tucker Loane Farrell, "Dr. Grover Tells His Tale," *Orlando Evening Star*, June 30, 1961.

Theo's variable experiences of the cut-flower business and his admission that it probably needed more perseverance are in Mead, *Autobiography*, 11. Vaughan's letter informing him that his terms were not commercially practical is from Vaughan Seed Store to Mead, June 9, 1923, RC-TLM.

CHAPTER 31: Bromeliads, Crinum & other Flowering Bulbs

Nehrling's view that Mead was a great hybridizer comes from Nehrling to TLM, January 7, 1924, RC-HN, and his comparison with Burbank is from *Nehrling's Early Florida Gardens*, 72.

The record of Theo's first bromeliad cross is from his notebook in MAS-TLM, Box 1, folder 1. It spans 45 pages of bromeliad activity in the 1922 to 1936 period, representing hundreds of crosses, including more than 40 different *Billbergia* hybrids as well as many other genera.

A hand-colored photographic print of this bigeneric cross, taken by Mead in 1928, and entitled "Billcrypta nutans–Beuckeri" resides in RC-TLM. The name is not accepted today since it uses a combination of the first parts of the genera involved.

According to Theo's diaries of the 20s and 30s (WC), Mulford Foster visited him at Lake Charm and received generous gifts of plants. On April 26, 1931, he spent four hours there and Theo gave him "lots of things," and on June 11, 1932, Theo records giving him "many plants." Theo sent him some *Gloriosa* lilies on December 14, 1931. Foster regularly listed Mead's "Cryptanthus-Billbergia" hybrid in his catalogs, e.g. Catalog No. 3, 1949/1950, MAS-BR.

Theo's desire to hybridize the finest sorts of iris comes from TLM to parents, March 19, 1889, RC-TLM. The description of the flowers of the Coral Lily as being "made of sealing wax" appears in TLM to parents, April 14, 1889, RC-TLM.

Receiving a large number of crinum from an English collector in India is referred to in Mead, *Autobiography*, 12. The preparation, planting, and subsequent hybridization are documented in Theodore L. Mead, "A Garden of Crinums," *The American Garden* 22:10 (1891): 597–598. C. Kircape appears in E. O. Orpet, "Three Good Plants," *Garden and Forest* 8:386 (1895): 288. Theo's only named crinum hybrid was *Peachblow*; Ogden, *Garden Bulbs for the South*, 146.

Originated in 1905, *Golden Measure* was a pure yellow gladiolus, a color long thought unattainable, and a rarity commanding a high price of $5 for a single bulb; Mead, *Autobiography*, 12. His patient approach to hybridizing is described in Theodore L. Mead, "Breeding for Novelties," *The Modern Gladiolus Grower* 3:6 (1916): 83. Circumventing the unwanted attention of hummingbirds comes from Theodore L. Mead, "Gladioli from Seed," *The Modern Gladiolus Grower* 2:6 (1915): 81. His disappointment with the Deland dealer is from Mead, *Autobiography*, 13.

Theo contribution to daylily hybridization is covered in Wyndham Hayward, "The Daylily in Florida," *Proceedings of the Florida State Horticultural Society* 63 (1950): 195. In 1941, Dr. Hamilton P. Traub hybridized a saffron yellow daylily and named it *Theodore Mead* in his honor.

The story of the rare *Hymenocallis* bulb eaten by a lubber grasshopper comes from Mead, *Autobiography*, 14.

The history of paperwhite narcissus growing in Florida is covered in Theodore Stephenson, Ouida Trammell Stephenson and John C. Van Beck, "Paper White Farming in Florida, 1928–1942," *The Daffodil Journal* 31:3 (1995): 162–165. Theo's diary of October 5, 1926, WC, records the planting of 40,000 narcissus bulbs at his Sanford farm.

Lucius D. Drewry was a successful insurance agent for the Mutual Benefit Life Insurance Company of Cincinnati and spent the winters in Florida, where he invested in flowering bulb production for the Northern markets. Details of the transactions are taken from Mead, *Autobiography*, 12, and letters TLM to Drewry, May 20, 1927, RC-TLM and Drewry to TLM, December 3, 1927, RC-TLM.

CHAPTER 32: The Oviedo Boys & Scoutmaster Mead

Following Dorothy's death, Theo expressed his desire for children as the only thing that would make life worth living in a letter to his parents, March 6, 1892, RC-TLM.

Theo was close to Edith's family at Coalburg and he became "Uncle Teddy" to the young children. Catherine Tappan Smith had four children, Charlotte Washington Willis (b. 1905), Katherine Anne Edwards Willis (b. 1906), John Augustine Willis, Jr (b. 1909) and Ogden Edwards Willis (b. 1916). Eleanor Dudley Smith also had four children, Julius Allen deGruyter, Jr (b. 1917), Anne Edwards deGruyter (b. 1922), Catherine Dudley deGruyter (b. 1924) and Elizabeth Stuart deGruyter (b. 1926).

"Never had a boy of my own" is a lament in TLM to Clarence Gilbert, October 27, 1915, RC-TLM. Wood shavings and a speck of potassium as a favorite pyrotechnic when the boys came round is from TLM to parents, May 17, 1897, RC-TLM. The water fight is described in TLM to Clarence Gilbert, 1915, RC-TLM.

The ink company, Gobol, marketed "Gobolinks" as the "goblins of the ink-bottle." They were produced by dropping a little ink on a sheet of white paper, folding the sheet in the center and pressing the ink spots together with the fingers. The pastime was enshrined in the book Ruth McEnery Stuart and Albert Bigelow Paine, *Gobolinks, or Shadow-Pictures for Young and Old* (New York: The Century Co., 1896). The construction of the 'Wondergraph' is given in F. E. Tuck, "How to Make a Wondergraph," in *The Boy Mechanic Book 1* (Chicago: Popular Mechanics Press, 1913), 436–439. Bending glass tubes over an alcohol lamp is mentioned in Theo's diary of August 6, 1911, WC.

The meeting notes of the Boy Scouts are from January 20, 1921, RC-TLM. At Christmas 1925, Scout diaries were presented to: J. B. Jones, Malcolm Jones, Bennie Jones, Arthur Partin, Tippy Partin, James Partin, Allen Thompson, Walter Carter, John Lawton, Jos. Lawton, Emmet Kelsey, Harvey Kelsey, Bernie Polson,

Jack Varn, J. W. Shuman, Bennie Wainright, Milton Gore, Warren McCall, Edward McCall, Olin Wright, H. B. Coney, A. D. Sauer, Hugh Kennedy, J. W. McNeal, Owen George, James George, Killis Booker, and Billy Brown. This list of recipients is from a single sheet manuscript in RC-TLM.

The "Moses or Santa Claus" comment is contained in Judi Grove, "Mead Garden Donor Created Small Eden," *Orlando-land*, August 1979, S20–21.

"Santa Claus just seemed to fit him," is from Claire Evans "The Happy Botanist and his Wife," in *It's About Time, Reflections from Central Florida*, 1:1 (2003): 20–23. Published by the Orange County Regional History Center, Orlando, April 2003.

The quote that includes the sense of "frozen horror" for Theo appears in the article by Edwin Osgood Grover, "The Making of a Botanical Garden," *Parks & Recreation* (1948): 452, and in the transcript of the speech Lewis Knudson gave at the dedication of Mead Garden in 1940 (RC-TLM).

John Hurd Connery was born on October 23, 1908, in Chattanooga, TN., and died on February 18, 1982, in Orlando, FL. He was a special student at Rollins College, Winter Park, FL., between 1931 and 1933.

Theo's birthday present of the second-hand Model T Ford touring car is from TLM diary entry February 23, 1923, WC. The citations for Theo's driving incidents are: "Must practice in safe area" (TLM diary March 10, 1923, WC); "Crashing through back wall" (Adicks and Neely, *Oviedo*, 51-52); "Sideswiped car returning from Sanford" (TLM diary November 8, 1924, WC); "Got as tired cranking" (TLM diary December 31, 1923, WC); "Keep a whiskbroom in this car" (Evans, *Happy Botanist*, 21).

A favorite lake trip with his scouts was to Crystal Lake, Silver Lake and Palm Springs. These three lakes, located north of Orlando in Longwood, became part of the Sanlando Springs resort complex in 1930 and became a major attraction as one of Central Florida's favorite swimming holes for many years.

Sweetwater Park, generated out of five acres of land given by the Meads, became Oviedo's first recreational park (Adicks and Neely, *Oviedo*, 96).

The double exposure picture of the boy with amaryllis taken March 23 is either Tom Brown or Charlie West according to Theo's 1924 diary entry of that date (WC).

The Scouts as "truly my own children" appears in Mead, *Florida Trucker*, 4. Theo's experiences at Camp WeWa are recounted in TLM to Ogden Willis, July 12, 1928, WC. Resigning as Scoutmaster comes from TLM to Ogden Willis, November 16, 1930, WC. H. J. Laney was the Principal of the Oviedo School in Seminole County. Joe Leinhart as a future commercial grower is from Evans, *Happy Botanist*, 22.

CHAPTER 33: Modern Conveniences & Medfly Crisis

The description of the 1926 invitation to the Alpha Delta Phi initiation comes from Mead, *Autobiography*, 6, and TLM's diary entry of February 20, 1926, WC.

Details of the purchase of the second car are revealed in TLM's diary entry of January 21, 1927, WC. The report of Theo's class of '77 reunion at Ithaca appears in *The Sanford Times*, July 26, 1927.

The narrative of Edith's death in October 1927 is assembled from letters: TLM to Anne Smith, October 19; TLM to Catherine Willis, October 21; TLM to Anne Smith, October 23; TLM to Catherine Willis, October 25, and TLM's diary entries October 19 to November 9, 1927. All sources WC.

Edith's way of rewarding her piano-playing pupils through the gift of a singing canary is remembered in Evans, *Happy Botanist*, 22. Thousands of male Hartz Mountain Rollers were shipped from Germany to the USA in the late nineteenth and early twentieth centuries, commanding high prices for the very best singers.

The swimming pool at Sweetwater Park now lies buried under earth but was a great attraction in its day according to Jenny Andreasson, "Pool Buried under

Park," *Seminole Voice*, March 7, 2008. http://archive.seminolevoice.com/Seminole_Voice/article.asp?ID=197, [accessed 28 January 2016].

Without Edith, Theo was lost at Christmas, as he admitted in TLM to Catherine Willis, December 8, 1927, WC. His regret in not spending Christmas at Coalburg is contained in TLM to Catherine Willis, December 13, 1927, WC.

Theo's story of Clayton's mismanagement of car maintenance while he was in the hospital comes from TLM to Catherine Willis, November 7, 1927, WC. Stories of Theo's continuing misfortunes as a driver come from TLM to Gus Willis, March 4, 1929 ("orange truck"), TLM to Ogden Willis, August 13, 1933 ("big puddle"), and TLM diary entry October 21, 1929 ("stuck in the driveway"). All sources WC.

News of Drewry's death and the collection of the final settlement come from TLM to Catherine Willis, February 21, 1929, WC and TLM to Willis family, May 3, 1931, WC.

Bill Edwards was the only son of William Seymour and Hope Edwards, born in 1906. The report of the plane crash is from *The San Mateo Times*, California, June 25, 1930, 8. Anne's funeral notice appeared in the *Charleston Daily Mail*, September 10, 1930.

News of the excellent citrus harvest of 1928/29 is from Adicks and Neely, *Oviedo*, 94. The description of the initial discovery of the fruit fly to the spraying with baited insecticide is an amalgamation of material from the following sources: Richard A. Clark and Howard V. Weems, "Detection, Quarantine, and Eradication of Fruit Flies invading Florida", *Proceedings of the Florida State Horticultural Society* 102 (1989): 159–160; Philip J. Pauly, *Fruits and Plains: The Horticultural Transformation of America* (Cambridge, MA: Harvard University Press, 2008), 224–229; and Ed L. Ayers, "The Two Medfly Eradication programs in Florida," *Proceedings of the Florida State Horticultural Society* 70 (1957): 67–69.

The first of many letters Theo wrote to the newspapers about the Medfly problem is from *The Sanford Herald* issue of May 15, 1929. The "Swat It" cartoon appeared in *Florida Clearing House News*, Winter Haven, Florida, Vol 1, No. 16, May 25, 1929, 1. The quarantine map is from Mark Carlson, Kris James Mitchener and Gary Richardson, "Arresting Banking Panics: Fed Liquidity Provision and the Forgotten Panic of 1929," National Bureau of Economic Research, Working Paper No. 16460, 2010. http://www.nber.org/papers/w16460, [accessed 28 January 2016].

The progress of the fly along the roads of travel is reported in "Florida Determined to Eradicate Pest; 2,900 Fighting Fly," *Florida Clearing House News*, Winter Haven, Florida, Vol 1, No. 16, May 25, 1929, 2. The roadblock quarantine confiscating tomatoes is from Malcolm Johnson, "Medfly an alarm word," *Ocala Star-Banner*, Oct 27, 1975, 7.

"Banks exploding like firecrackers" appears in TLM to Gus Willis, July 22, 1929, WC. Theo's attempts to protect his precious tropical plantings by putting up signs come from TLM's diary, August 25 and 27, 1929, WC. "Destroy all our industries" is taken from TLM to Willis family, March 3, 1930, WC.

The final eradication cost is found in Clark and Weems, *Proceedings of the Florida State Horticultural Society*, 160. Newell's legacy is examined in Laurence A. Mound, "Florida Pioneer Wilmon Newell: The Past, Present and Future of Insect Pest Control," *Florida Entomologist* 88:2 (2005): 241–243.

CHAPTER 34: Awards, Recognition & Culmination

The Florida Garden Club's tribute to Theo appears in *The Florida Times-Union*, January 16, 1927. Details of the Arbor Day 1931 celebrations are from *The Central Florida Press*, Oviedo, Florida, January 16, 1931, and TLM to Catherine Willis, January 11, 1931, WC. "Everybody very lovely to me" is a TLM diary entry from January 9, 1931, WC.

Theo's eulogy to Henry Nehrling is taken from *The Orlando Morning Sentinel* of November 30, 1931. Theo's experiences at the Alpha Delta Phi centenary celebrations come from *The New Yorker*, September 17, 1932, 32 and TLM's diary of September 4, 1932, WC.

Theo's discussion on the hybridization of amaryllis is announced in *The Palm Beach Post*, October 14, 1929, 2. The *Orlando Sunday Sentinel*, March 23, 1930, reports the amaryllis *T. L. Mead* being shown at the Orlando Flower Show. News of the sale of his entire collection of bulbs of America's first white amaryllis is from I. W. Heaton to TLM, August 15, 1935, RC-TLM. The Heaton Bulb and Palm Co., Orlando were agents for Theo's bulbs at that time.

The comment on the challenge of growing and hybridizing orchids is contained in the letter he sent to one of his old German school friends, Tobias, January 30, 1903, RC-TLM. Turning down $500 for a rare orchid originates in Eastman, *Southern Personalities*, 7. Theo's prose mastery regarding orchids appears in T. L. Mead, "Orchids – Interesting Facts as to their Beauty, Habits and Culture. List of Varieties Best Adapted to Culture in Florida," *Proceedings of the Florida State Horticultural Society* 10 (1897): 38–50.

The Mead collection of bromeliads passed first to Harry Smith (Terry Bailey, "Gotha's Nally Estate: A Tropical Paradise," *Winter Garden Times*, May 24, 1979), then to Julian Nally at Gotha (Julian Nally, "Bromeliads by the Acre," in *Bromeliads, A Cultural Handbook* (The Bromeliad Society, 1953), 50). Mead as the first person to hybridize the bromeliad in North America is documented in Victoria Padilla, "Bromeliads in American Horticulture," *Journal of the Bromeliad Society* 35:1 (1985): 5. Theo's brief summary of his time hybridizing bromeliads is from Mead, *Autobiography*, 13.

The comment concerning Theo's garden visitors is taken from TLM to Eleanor deGruyter, March 7, 1931, WC.

In 1930, as official photographer to the Beebe expedition, Jack Connery recorded the events of the first record-breaking dive to 803 feet and the second, on June 11, to 1,426 feet. Unfortunately, that summer, he slipped between the *Skink*, Beebe's twenty-six-foot power launch and the wharf, fracturing a bone in his back and had to abandon the expedition; reference Brad Matsen, *Descent: The Heroic Discovery of the Abyss* (New York: Pantheon Books, 2006), 102. Connery's agreement with Rollins College is stated in Farrell, "Dr. Grover Tells His Tale."

Connery's involvement with the scouts and orchid growing come from Theo's 1932 diary, WC: April 15 ("brought Rollins students"); April 18 ("projector entertainment to scouts on birds"); and July 22 ("digging caladium and repotting orchids").

Theo's surprise birthday party at the Connery's in 1933 is described in TLM to Willis family, February 24, 1933, WC.

John Hurd Connery and Helen Corey Golloway were married on May 30, 1934, in North Canton, OH. News of Helen's subsequent appointment as personal secretary to Dr. Edwin Grover is contained in a letter she sent to Donna Rhein, at the Winter Park Library, on July 15, 1995, WPPL.

Theo's initial trip to the dentist in 1929 is disclosed in TLM to Gus Willis, April 8, 1929, WC. His difficulty in hearing became noticeable when listening to sermons in church, as he reported in TLM to Catherine Willis, December 1, 1930, WC. Despite this hearing trouble, the letter goes on to say that in the afternoon, after church in the morning, he went to see the movie *Tom Sawyer* with Jackie Coogan (a talkie) and enjoyed it immensely.

Theo's report of trading in his old Victrola is from diary entry October 28, 1930, WC, and his preference in music in TLM to Gus Willis, December 4, 1929, WC.

"Always looked that way when you had had your teeth out" is in TLM to Catherine Willis, November 1, 1931, WC. Theo's description by the minister and the "amiable cuss" quote is recorded in TLM to Catherine Willis, August 2, 1931, WC. Replacing his hearing horn with a Sonotone hearing aid is from TLM to Catherine Willis, October 13, 1933, WC. Struggling to entertain his Oviedo boys is a confession made in TLM to Gus Willis, June 2, 1934, WC.

"Not much fun if a fellow can't hear" comes from TLM to Ogden Willis, June 8, 1935, WC. His reply to the Alpha Delta Phi invitation is reported in "Mead '77 Regrets Missing Initiation; Writes a Tribute," *The Alpha Delt* 1:1 (June 1935), 4.

An asset list, with Clayton's mortgage and celery plot cost, dated October 1,

1929, is in RC-TLM; the decision to donate the flivver to Clayton is taken from TLM to Miner Berry, July 6, 1934, WC.

The Florida Federation of Women's Clubs established Royal Palm State Park, a large hardwood hammock containing many Royal Palms, in 1916. When it opened, Theo made a gift of a large quantity of his orchids and a section of the park became the "Mead Orchid Garden." In the 1940s, the Federation gave the land to the federal government and it became part of the Everglades National Park in 1947.

Theo had a special relationship with Rollins College, visiting Winter Park frequently and attending several Rollins events from the mid-1890s onward. He met the new president of Rollins College, Charles Fairchild, in 1894 and was Special Lecturer in Botany there from 1896–1901.

A copy of TLM's will, dated August 19, 1933, with a sentence canceling the indebtedness for Clayton added as a postscript is in RC-TLM.

How Theo came to change his mind over the fate of his orchids is contained in Mead, *Autobiography*, 11, and in Jack Connery to Clifford Cole, January 15, 1935, RC-TLM. Theo's promise of a share of orchids to Connery is from TLM to Gus Willis, December 17, 1934, WC.

E. C. Cole was the president of the Cole Manufacturing Company of Chicago, known for "Cole's Hot Blast Stoves." In 1922 his son, Clifford built the Walter De Garmo designed Villa Mira Mare on the bay front in Coconut Grove, at 2484 S. Bayshore Drive, now the Coral Reef Yacht Club.

The reason for Theo's confusion is unknown, but it is possible that bacterial infection from a decayed tooth was the cause. His high blood pressure was a further concern as detailed in Mrs. Berry to Catherine Willis, January 15, 1934, WC, but by March had returned to normal (TLM to Willis family, March 10, 1934, WC). Both Mrs. Berry (letter to Willis family, May 31, 1934, WC) and Theo himself (letter to Gus Willis, April 25, 1934, WC), noted the differences in appearance and mental faculties following the infection.

Theo's philosophical attitude to his things been organized under Miss Dixon is from TLM to Gus Willis, February 26, 1934, WC. He found it harder to accept when she tried to organize him, as he confessed in TLM to Gus Willis, June 2, 1934, WC. This prompted the question of a new housekeeper, suggested in Gus Willis to Mr. Berry, December 13, 1934, WC. Mrs. Carter's choice of Mrs. Wainright was a relief to the Willis family—Gus Willis to Mrs. Carter, December 20, 1935, WC.

The overgrown appearance of *Waitabit*, and "Peter Pan and the joy of living" come from Nina Oliver Dean, "Candlelight," *The Sanford Herald*, Saturday, August 29, 1931. A similar first impression of Theo's cottage and entrance appears in Eastman, *Southern Personalities*, 7.

"Brotherhood and fraternal love" come from Mead, *Autobiography*, 5.

The request from Nally for a visit to see Theo is contained in Julian Nally to TLM, December 14, 1935, RC-TLM. Nally's account of the visit is from a draft partially completed manuscript in MAS-JN, Series II: Personal, 1902-2001. Nally's official response following the visit is taken from Nally to TLM, undated, 1936, RC-TLM.

Theo's last hybridization success is documented in his notebook for April 14, 1936, MAS-TLM. Details of his funeral service are in an "In Memoriam" booklet, WC.

CHAPTER 35: Aftermath & Legacy

"One of the world's most distinguished entomologists and horticulturists" comes from the *Orlando Morning Sentinel* of May 5, 1936. Julian Nally's recollection is from an undated draft manuscript in MAS-TLM.

Mead's contribution to bromeliad culture is reported in Paul Butler, "Revisiting T. L. Mead's Contribution to Early American Bromeliad Hybridization," *Journal of the Bromeliad Society* 62:5 (2012): 214–223.

Mead's contribution to orchid culture has been celebrated in Edwin Osgood

Grover, "Theodore Luqueer Mead," *The Orchid Journal* 2:4 (1953): 159–163, and in Paul Butler, "Theodore L. Mead: Pioneer American Orchid Grower & Hybridizer," *Orchids* (Published by American Orchid Society, Jan 2014): 34–37.

The Herbert Medal was named after William Herbert (1778–1847), a British botanist, best known for his early taxonomy of bulbous plants.

Connery's promise to build a memorial garden is referred to in Grover, "The Making of a Botanical Garden", 452. Grover's statement that every garden in Florida had at least one 'Mead' plant is contained in the transcript of his remarks at the Mead Garden groundbreaking ceremony in January 1938, RC-TLM.

The story of the creation of Mead Botanical Garden and its early history is covered in a number of newspaper articles, for example, The Miami Daily News, March 6, 1938, 15, as well as Grover's "The Making of a Botanical Garden." The Florida tourist attraction map is taken from Tim Hollis, *Selling the Sunshine State* (Gainesville: University Press of Florida, 2008).

Henry Nehrling's earlier tribute to Theo appears in Nehrling's *Early Florida Gardens*, 72. The Florida State Horticultural Society's tribute can be found in *Proceedings of the Florida State Horticultural Society* 39 (1936): 159. Mead's downplaying his skill of hybridizing comes from Eastman, *Southern Personalities*, 42.

"Love seems to be the only thing…" is revealed in a letter to his parents, April 20, 1895, RC-TLM.

Picture Credits

Front Cover: Detail from "Orchid and Hummingbird near a Mountain Waterfall" with addition of detail from "Blue Morpho Butterfly." Both images in the public domain and from original oil paintings by Martin Johnson Heade (1819–1904).

Frontispiece: Courtesy of the Willis family.

1.1L: From "New York's Oldest Grocery," Duluth Evening Herald, Tuesday, June 28, 1892, 34.

1.1R, 5.4, 5.5, 6.4, 9.4L, 10.2, 12.3, 13.2, 14.1R, 14.3R, 16.3, 18.3, 19.1, 21.1, 22.3R, 23.3, 24.1, 25.1, 26.1, 27.1, 27.2, 29.4, 30.1, 31.1, 32.1L, 32.7, 32.8, 33.1, 34.2R: Used with permission, Department of College Archives and Special Collections, Olin Library, Rollins College, Winter Park, Florida.

1.2: From *Our Excellent Women of the United Methodist Church*, (New York: James Miller, 1873), 221.

1.3, 2.2, 3.1, 3.3, 5.1, 6.2, 6.3, 6.5, 8.1, 8.2, 9.3, 9.4R, 10.1, 12.1R, 12.4, 14.1L, 14.4, 15.2, 16.1, 18.1, 18.4, 19.3, 20.1, 20.2, 20.3, 21.2, 22.4, 23.2, 24.2, 25.3, 26.2, 27.3, 29.2, 30.3, 30.6, 32.2L, 32.4, 32.5, 32.6, 34.2L, 34.3, 34.5, 34.7: Courtesy of the Willis family.

2.1: https://commons.wikimedia.org/wiki/File: Cecropia_Moth_Caterpillar_(Hyalophora_cecropia).jpg, [accessed 14 January 2016].

3.2: http://www.wikiwand.com/de/Bad_Homburg_vor_der_Höhe, [accessed 14 January 2016].

4.1: https://en.wikipedia.org/wiki/Otto_Staudinger, [accessed 28 January 2016].

5.2: From *Whitney's Florida Pathfinder for the Tourist and Invalid,* (New York: John Prescott Whitney, 1876). Map of the Upper St. Johns River from Palatka to Enterprise, hand-colored by Lisa Middleton at greatriverarts.com. Used with permission.

5.3: Drawing of alligator shooting from a steamer on Lake Ocklawaha. ca 1874. Black & white photoprint, 8 x 10 in. State Archives of Florida, Florida Memory. https://www.floridamemory.com/items/show/25950, [accessed 19 February 2016].

5.6, 7.3: From William H. Edwards *The Butterflies of North America*, (Vol. 1, Philadelphia: Amer. Entomol. Soc., 1868–1872; Vol. 2, Boston: Hurd & Houghton; Houghton, Osgood & Co.; Houghton, Mifflin & Co., 1874-1884; Vol. 3, Boston: Houghton, Mifflin & Co., 1887–1897).

6.1: From National Geographic's MapMaker Interactive.

7.1L: From Andrew Morrison *The City of Denver and State of Colorado*, (Engelhardt Series, 1890), 22.

7.1R: From *The Life of Hon. William F. Cody,* (Hartford: Frank Bliss, 1879), preface.

7.2: https://en.wikipedia.org/wiki/Mount_of_the_Holy_Cross#/media/File: Mountain_of_the_Holy_Cross,_Colorado_-_NARA_-_517691.jpg, [accessed 28 January 2016].

7.4L, 7.5, 8.3, 14.3L, 17.2, 34.9, 35.1: From author's collection.

7.4R: Used with permission of the Museum of Comparative Zoology, Harvard University.

9.1: From Theodore L. Mead, An Autobiography, *The Yearbook of the Amaryllis Society,* (1935), 2: 5.

9.2: Photographed by Beardsley, Ithaca, 1875, Biographical Collection, History Center, Winter Park Public Library, Winter Park, FL.

11.1: https://en.wikipedia.org/wiki/File:SierraMadreVillaHotel-1884.jpg, [accessed 28 January 2016].

11.2: https://commons.wikimedia.org/wiki/File: Tourists_on_horseback,_Yosemite_Falls,_California,_from_Robert_N._Dennis_collection_of_stereoscopic_views.jpg, [accessed 28 January 2016].

12.1L, 12.2: Used with special permission from Alpha Delta Phi at Cornell University.

13.1: From H. K. Ingram *Tourists and Settlers Guide to Florida,* (Jacksonville: Da Costa Publishing, 1895-96), preface.

13.3: From *Illustrated Florida*, (Buffalo, NY: Dodge Art Publishing, 1882), 10.

13.4: Redrawn from 1890 Plat, S H Mead Estates, Book 1, Page 6, Instrument no. PB001P006-1, Record date 6/2/1890. http://officialrecords.lakecountyclerk.org, [accessed 20 December 2015].

14.2: Courtesy of Betsy Rose.

14.5: Used with permission of the West Virginia Archives & History Library.

15.1: From *The Illustrated London News*, August 7, 1852, 97.

16.2L, 16.2R: Hein Nouwens/Shutterstock.com; Dn Br/Shutterstock.com.

17.1: From Samuel Hawley Adams *Life of Henry Foster, M. D.*, (Rochester, NY: Rochester Times-Union, 1921), 43 & 61.

17.3: Steam launch on Lake Eustis - Lake County, Florida. 192-. Black & white photonegative, 4 x 5 in. State Archives of Florida, Florida Memory. https://www.floridamemory.com/items/show/143174, [accessed 19 February 2016].

17.4: Memorial church and parsonage - Lake Charm, Florida. 189-. Black & white photoprint, 5 x 8 in. State Archives of Florida, Florida Memory. https://www.floridamemory.com/items/show/26102, [accessed 19 February 2016].

18.2, 19.2, 22.1, 22.3L, 26.3, 26.4, 26.6, 28.1, 28.2, 30.2, 31.2, 32.1R, 34.6: Photography by Theodore Mead, Helen Connery Collection, History Center, Winter Park Public Library, Winter Park, FL.

22.2: From Country Life in America, 7 (1905): 368.

23.1: Postcard published by M. Mark of Jacksonville, FL., circa 1908-1913.

25.2: Courtesy of the archives of The Cathedral Church of St Luke, Orlando.

26.5: From Thomas Condit Miller and Hu Maxwell *West Virginia and its People Volume* 3, (New York: Lewis Historical Publishing, 1913), 842.

29.1: Courtesy of the Nehrling Gardens archives, Gotha, FL.

29.3: Robinson, T. P., b. 1870. *Professor Henry Nehrling in his garden in Gotha, Florida.* 1908. Black & white photoprint, 7 x 9 in. State Archives of Florida, Florida Memory. https://www.floridamemory.com/items/show/257876, [accessed 19 February 2016].

30.4, 30.5, 32.3, 34.4, 34.8, 35.3: Courtesy of The Winter Park Garden Club, Winter Park, FL.

32.2R: http://www.legrenierdepascal.com/2011/12/wondergraph.html, [accessed 31 January 2016].

33.2L: From *Florida Clearing House News*, Winter Haven, Florida, May 25, 1929. Vol 1, No. 16.

33.2R: From Mark Carlson, Kris James Mitchener and Gary Richardson,

"Arresting Banking Panics: Fed Liquidity Provision and the Forgotten Panic of 1929," National Bureau of Economic Research, Working Paper No. 16460, 2010. http://www.nber.org/papers/w16460, [accessed 28 January 2016].

34.1: Used with permission of Fairchild Tropical Botanic Garden, Miami, FL.

35.2: Derivative image created from original prints of Connery (left), courtesy of The Winter Park Garden Club, and Grover (right), used with permission, Department of College Archives and Special Collections, Olin Library, Rollins College, Winter Park, Florida.

35.4: From *Scenic Florida*, (Tallahassee, FL: Florida State Department of Agriculture, circa 1940), 19.

Final adieu sign off image: Used with permission, Department of College Archives and Special Collections, Olin Library, Rollins College, Winter Park, Florida.

Acknowledgement image: Author's collection.

www.ingramcontent.com/pod-product-compliance
Lightning Source LLC
Chambersburg PA
CBHW061924290426
44113CB00024B/2818